MINDFUL SCEPTICS
GUIDE TO...

growing enough food

Mark Dangerfield
Christopher M Scott

Mindful Sceptics

Published by Mindful Sceptics

Mindful Sceptics

Springwood, NSW 2777, Australia

mindfulsceptics.info

AI Use Statement

This book was conceived, written and edited by a human who occasionally used AI tools to assist with grammar, editing, summaries and focused searches.

Contents

Preface to the Series

Ever feel like you're drowning in information and controversy, unsure what to believe? You're not alone. In this age of information overload, finding clarity can seem like searching for a needle in a digital haystack.

But what if I told you that the key to navigating this chaos lies within you?

Welcome to the Mindful Sceptic Guides, your compass in the turbulent waters of modern discourse. These aren't your average self-help books or dry academic tomes. They're intellectual adventures designed to awaken your innate curiosity and sharpen your critical thinking skills.

Imagine being able to cut through the noise, to see past the spin and the hype. Picture yourself in conversations where your insights turn heads, where your words carry the weight of well-reasoned thought. That's the power of being a mindful sceptic, a power that's within your reach.

In the pages of *The Mindful Sceptic Guides*, you'll find a delightfully eclectic mix of crucial facts, thought-provoking ideas, and paradigm-shifting perspectives. We'll tackle big issues with a unique blend of scientific rigour and philosophical inquiry. From conservation to social justice, from food security to healthy diet, economic theory to personal growth, no topic is off-limits.

But here's the kicker. This isn't just about personal enlightenment. The stakes are higher than that. In a world grappling with complex, existential challenges, humanity's survival may depend on cultivating more mindful sceptics—people who can question dogma, seek evidence, and imagine new solutions to age-old problems.

Could you be one of these crucial thinkers? Could your ideas be the spark that ignites positive change?

There's only one way to find out. Dive into these guides with an open mind and a willingness to challenge your assumptions. Prepare for moments of revelation, ideas that will keep you up at night, and insights that will transform how you see the world and your place in it.

Are you ready to embark on this intellectual journey? To become a beacon of reason in a world that desperately needs it? Then, turn the page. Your adventure in mindful scepticism begins now.

Welcome to the Mindful Sceptic Guides. Let's change the world, one critical thought at a time.

Preface to this Guide

Imagine a world where every grocery store shelf is empty, where the hum of refrigerators is replaced by the growl of hungry stomachs. Now, take a deep breath and look around. We're not there yet, and that's nothing short of a daily miracle.

Every 24 hours, humanity pulls off a feat that would make even the most ambitious logistics expert's head spin. We produce, distribute, and consume a staggering 22 trillion kilocalories of food. That's not just a number—it's breakfast, lunch, and dinner for 8 billion people, from the bustling streets of Tokyo to the remote villages of the Amazon.

But here's the kicker. We can't afford to miss a beat. One hiccup in this intricate dance of agriculture, transportation, and commerce is not just about disappointed shoppers facing an empty shelf. We're talking about a domino effect that could topple societies.

Welcome to *The Mindful Sceptic Guide to Growing Enough Food*, where we dive headfirst into the most fundamental questions. Can we keep this miracle going? Can we feed a growing population of humans on a planet with finite resources?

This isn't just another dry analysis of crop yields and fertiliser ratios. We will journey to the foundation of human civilisation—our ability to feed ourselves. We'll explore the triumphs of modern agriculture, the looming challenges on the horizon, and the innovative solutions that could revolutionise how we think about food production.

We'll challenge assumptions, bust myths, and ruffle a few feathers. Because when it comes to feeding the world, there's no room for complacency or magical thinking.

Whether you're a farmer with soil under your nails, a policymaker grappling with food security, or simply someone who enjoys a good meal, this guide has something to offer you. Prepare to have your perspective broadened, your assumptions challenged, and your role in this global food web illuminated.

So, are you ready to become a mindful sceptic about food production? To see beyond your plate and into the complex, fascinating, and crucial world of growing enough food for humanity? Then, turn the page. Your next meal and the future of human sustenance awaits.

Chapter One

Global food production and sustainability

Profit is measured in dollars, not surplus. Consequently, a profit imperative would still prioritise yield gain over resilience because it speeds up the conversion of natural capital to cash—you cannot buy a new car with ten years worth of soil carbon reserves.

Sustainable food is essential

Let's go on a food journey not of culinary delights but of what it takes to grow, harvest, store and distribute enough food for 8 billion people and their pets—a story that will leak mindfulness into your day.

All humans have to eat. We can only exist without food for a few weeks before our bodies begin to eat themselves. Because of this necessity, what we eat, when and how has always been central to individuals, groups and societies.

Our success as a species owes much to our collaborative skills in finding, storing, cooking and sharing food from almost every place on Earth. Early humans learned to eat whatever was edible in each habitat. They used their control of fire to cook and large brains to communicate and pass on knowledge of how to find and prepare edible foods.

Then, we made a breakthrough by figuring out how to grow and rear food. The invention of agriculture just 12,000 years ago is staggeringly recent in the 300,000-year history of *Homo sapiens* and just a blink in the 6 to 7 million years since the first ancestral *Hominid*, but it meant we could persist in large groups, adopt a division of labour and create cultures with unimaginable technologies and complexity. It also allowed people to become more numerous as farming technologies spread across the planet.

Then everything changed.

It is tempting to think that the invention of agriculture was pivotal in human history. It was a big deal that changed many things but wasn't the most significant change. The big pivot came when humans made agriculture an energy sink.

Arguably, two features of the Industrial Revolution, an adoption and an invention, changed agriculture. Humans have known about coal, oil and gas for a long time. We had electricity, steam engines, railways, and steamships that made a difference in food production. However, the commercial extraction and use of oil in combustion engines from the early 1900s—the adoption—and the Haber-Bosch process to make fertiliser on an industrial scale using natural gas—the invention—made food plentiful. Oil and gas applied to agriculture was the change.

Almost overnight, farmers had machines to provide the labour to cut down vegetation, plough soil, fertilise fields and transport products to markets far and wide. They also had synthetic fertilisers to provide the nutrients. Agriculture went from an energy source to an energy sink and a food production powerhouse. Today, food production is staggering across a dizzying array of commodities and volumes—2.7 billion tonnes of grains, 342 million tonnes of meat, 154 million tonnes of fish—that are hard to picture. Dump just one tonne of grain in your living room, and there is no room for the cat.

The addition of machines, fertilisers, and pesticide inputs is called the intensification of agriculture. It increases yields and allows for economies of scale, but it turns food production into a net energy user and a business. Global food production generates trillions of dollars in revenue because two out of three people buy at least some of their food from a store or a vendor. About a third of the global population still grows most of their food, but for over 5 billion people, their food supply comes via a marketplace.

In a few generations, not even a blink in evolutionary time, modern humans created a technologically sophisticated, six-continent, just-in-time food supply chain that could feed everyone.

Large agricultural enterprises are profitable

A collective paradigm runs Western economies. We, the people, consume. It has been this way since the first agricultural systems created an energy surplus by generating more food energy than the effort it took to produce the crops and livestock. Over time, more and more people buy from the energy surplus generated not by people alone but by people aided by livestock and, since the 1900s, an army of fossil fuel helpers.

In modern times, over half of humanity lives in urban areas and has little choice but to buy food from the food energy surplus. A buying public is excellent for the businessman who can mobilise capital and labour to provide the goods that people need for less than the cost of producing them. A positive balance sheet generates a monetary surplus (profit) as a fungible asset (cash) that can quickly become more capital to invest. This is the positive feedback cycle we call capitalism.

Capitalism has ensnared global food production, the six-continent food supply chain that feeds over half the people. Farm enterprises geared for profit look for efficiencies that reduce their operating costs or increase their net output. More extensive and uniform is typically better for efficiently using machinery and energy inputs. It shouldn't be surprising that intensification happens because it delivers higher yields and economies of scale. Surplus gains accrue with less cost. The focus on efficiency extends to externalities and loss of resilience outside the balance sheet to help make intensification profitable. But there are other motivations for farming.

Not all farmers are in it for profit

Globally, 500 million people are growing food for themselves and their families, just enough surplus energy for people to live in the same place year-round and meet their basic needs. These operations are generally small and use minimal inputs. Farmers leverage their ingenuity, imagination and hard work to grow enough food and cash crops to pay the bills and turn a small profit in good years. Constrained by their access to land, inputs and labour, many do this on a hectare or less; small plots scattered across the landscape.

The global challenge is that the modest surplus from these small farms cannot feed the urban populations—over 6 billion by 2050—who need food. The conventional wisdom is that most global food production will be grown in intensive commercial production systems that can generate enough surplus for the cities. It is a very tempting proposition for the capitalist.

Global food production surplus to 2050

The current world population of 8 billion is expected to reach 8.6 billion in 2030 and 9.8 billion by 2050, assuming that the global production and consumption systems persist through the population spike. We will leave the collapse scenarios where the systems break down for later and assume that the current trajectory persists.

A common assumption is that meeting the food energy and nutritional needs of so many people will mean closing yield gaps and increasing global crop yields by 70–110%. What is grown will also need to diversify to improve the nutrient content of foods and nutrient and water use efficiency. Agricultural production must also change from a greenhouse gas emitter to a carbon sink by managing inputs and sequestering carbon into the soil. The only known way for agriculture to do these things is to wean itself off fossil fuel inputs and become circular, where inputs to soil come from recycling unused biomass and animal waste. This means part of the surplus production goes to feed the soil.

Circular agriculture will reduce the need for inputs but, in many ways, is the opposite of intensification, which has been the source of increased yields over the last 100 years. Agriculture that recycles rather than throughput costs less but is less productive, at least initially, to compensate for slow recovery. This means an additional 30% more global food production, increasing the total required global crop yield by 2050 to 100–140%.

Calories are only part of the challenge. The food surplus needs to be nutritious, too. Already, 1.8 billion people suffer from nutrient deficiencies called hidden hunger—twice the number of people who do not get enough calories.

What nature does with surplus

Nature knows what to do with a surplus. It converts it into more nature. It should be no surprise that food production surpluses result in more people, and once these people are present, there is an imperative to keep them fed. Greater production creates a disaster waiting to happen unless that production is maintained.

The future of humanity will be determined by how much we understand the metabolic energy surplus that we generate from agriculture. Not enough surplus, and we risk collapse. Too much surplus and we make more people to pressure future supply. And the kicker is that too much of the wrong surplus and people are unhealthy and unhappy. In other words, global food production is more than yield.

We, the people, will consume. It is inevitable, and the capitalists are licking their lips. Fortunately, we, the people, have a will and are smart enough to realise that we cannot consume a non-existent or nutrient-hollow surplus.

Humanity needs a massive food surplus to feed growing urban populations encouraged by the capitalist model to consume. However, people do not need the ads and policy encouragement because they eat or starve.

Demand for food is unprecedented, and recent

The industrialisation or intensification of agriculture happened in a little over 100 years—the energy from fossil fuels went into food production and created a lot of food. As the farmers grew more food, the local and global trade flourished, and most of the time, most people in most places could find food. And when plenty of food is available consistently, biology happens. People ate this food and made more people.

Energy from fossils was converted into people currently at a rate of 8,000+ per hour. All of a sudden, billions of people each need over 2,000 kilocalories daily to power their bodies together with balanced nutrients to keep them healthy. This creates a 22 trillion kilocalories a day challenge, the food energy needed to keep 8 billion people alive. Today, the demand for food is unprecedented in human history.

No matter, people are tenacious and innovative and can sniff an opportunity from a thousand yards. Plus, there are fossil fuel slaves to help do the heavy work. Humans set about realising the opportunity offered by billions of consumers captured by the necessity of eating, and a global supply chain was built to profit from the chance.

Where does the food come from?

Food production happens on land and a little in the oceans, lakes and rivers. More specifically, it occurs in soil. Soil is where seeds germinate, roots propagate, and go on to exchange water and nutrients with the growing plant. In the soil are billions of organisms of millions of types assisting with the nutrient and energy transfer to the plant. Almost all the food farmers grow depends on soil and the soil biodiversity that keeps soil healthy. Ask any farmer how easy it is to produce a crop in inert dirt, and he will laugh at you.

In the soil, the plants grow, capturing the sun's energy via photosynthesis to reproduce. They must survive against competitors, disease, the weather, and any number of animals looking to eat them, but nature has equipped them well, and most of them make more, most of the time.

Agriculture is the channelling of what plants and animals do naturally—more making—into the parts of plants and animals that humans can eat. We consume the nutritious parts of the plant: the seeds, fruits, kernels, tubers and fleshy leaves. Then, we feed the plants and plant parts that we cannot eat to animals, which converts the inedible parts into animal protein that is highly nutritious for humans.

People are good at animal husbandry, choosing the animal species that produce well and yet are easily controlled. After a few thousand years of domestication, livestock makes up 62% of the mammal biomass on earth today, the animals humans manage for our benefit. Animal protein is so important to us that we feed crops to cattle, sheep, goats and pigs. The 7 billion livestock animals in the United States consume five times as much grain as is consumed directly by the entire American population. More remarkable still than these feats of dominion is that collectively, food supply, more or less, matches demand.

Food production happens on farms

There are over 570 million farms worldwide. Farmers grow enough food on these farms to feed themselves, their livestock, and everyone else. Global food production is staggering. Farmers not only grow enough food to feed everyone; they grow enough food globally for humans to waste a third of the output. Losses at harvest, transport, storage, processing, preparation and on the plate are more than enough to feed another 2 billion people.

Equal to this miracle of the global food supply is how food is grown and reared. At least 2 billion people can feed themselves by growing their food. Five of every six farms of the 570 million farms are less than 2 ha in size. The farmers who work in these smallholdings produce around one-third of the world's crops, work hard and tend to get higher farm productivity but lower labour productivity from their land. Many of these small farms operate on limited external inputs and use techniques of production honed over millennia to suit local soils, climates and technologies. We call these systems subsistence farming with the unwarranted assumption that it is inferior to industrial production.

Then there are the 6 million farms greater than 50 ha in size. These are almost all high-input, intensive agriculture farms, the 1% of farms that operate more than 70% of the world's farmland. Food production from these large enterprises feeds the people in the cities. The UN estimates that by 2030, two out of three people, nearly 6 billion, will live in towns and cities. Most will have little or no opportunity to grow their food and rely on the complex, multi-continent food supply chain for sustenance.

Beyond these broad categories is layer after layer of nuance on what food can be grown, where and how.

Food supply chain

Growing food is just the start. Getting food from the paddock to the plate is a complex and complicated sequence of actions and transactions—the food supply chain.

At the beginning of the chain are the farms and farmers growing the crops and rearing livestock, and at the end are retailers selling products to consumers. The multitude of exchanges along this supply chain introduces food economics, where food becomes commodities, the financing of production and distribution, and the financial risk associated with a volatile production and market system.

Whilst there is money to be made in this most fundamental of transactions—remember everyone has to eat—the World Bank estimates the proportion of people involved in food production has declined globally from 44% as recently as 1991 to just 27% in 2019. In the OECD countries, this proportion is as low as 1%. As the human population grows, the proportion of people in agriculture declines. Proportional change is partly the consequence of large numbers, but the absolute number of agricultural workers has also declined. Farmers get the fossil fuel workforce to do all the heavy lifting and need fewer human workers. Infrastructure and technology reduce the need for intermediaries, especially as large corporations dominate the transactions along the supply chain.

The Pareto rule

This tale of agricultural production is not one of ever greater efficiency, tempting as that explanation might be. Wherever food production happens, a range of efficiencies typically follows the 80:20 rule—that 80% of production comes from 20% of producers.

Food production does not make a profit everywhere, even in the more intensive systems. There are many issues here, from investment, cost and farm debt to vested interests. It does not all smell like roast pork on a Sunday lunchtime, not least the higher risks of simplified yet intensive production.

People don't call it food security

The fact that billions of humans alive today have food is a miracle of human ingenuity and the generosity of nature. But for individuals, feeding yourself and your family takes effort and comes with risk. Food must be present nearby and preferably in your pantry every day. The food you have must make sense for your culture, culinary skills, and stomach. And there is no point in there being times of plenty if next week, month or year, food is in short supply. Secure food is nutritious, healthy, safe and reliable.

The World Bank estimates that although extreme poverty has declined, at least until the ravages of the COVID-19 pandemic, nearly half the world's population—over 3.5 billion people—struggle to meet basic needs of food, clean air, water and shelter. They survive on less than $5.50 per day.

It is not called food security by people who do not know where their next meal is coming from or might use over half their income to source food, but the concept can help us understand sustainable food. Food security has four components: availability, access, utilisation and stability.

Food must be present in your neighbourhood. You must be able to access that food by growing it or buying it. A kilo of rice is no help to you unless you have a stove and pan to cook it because the ability to utilise the food is just as crucial as having it. And then today's meal is of little help if the food runs out in the future for one of the first three reasons.

These four pillars of food security apply to both calories and nutrients. A plentiful supply of rice and nothing else can keep a person alive for a long time until they succumb to illnesses of malnourishment. Food security means enough food **and** nutrition for metabolic health. These ideas get a deeper dive in the *Mindful Sceptic Guide to Food Security*.

The global numbers associated with the absence of one or more of these pillars, food insecurity, are sobering.

- 9% of the world's population—700 million—has a caloric intake below minimum energy requirements.
- 663 million people globally are undernourished.
- 22% of children younger than five are 'stunted' due to poor nutrition or repeated infection.
- One in four people globally—1.9 billion—are moderately or severely food insecure.

Something is wrong if enough food is grown, but 700 million people are starving, malnourished and food insecure because one or more of the components of food security have failed. The truth is, on a global scale, we are not feeding everyone well. Abhorrent as this situation is, it is not the biggest problem. That dubious honour goes to the challenge presented by sustainable food.

What is sustainable food?

Sustainable food is what the national and global populations eat via the local and global food supply chain. It should be food that is available, edible, nutritious and consistent. The key word, though, is sustainable. Can this production persist indefinitely? And what happens if there are disruptions? And then there is the big question.

What happens when the fossil fuel energy subsidy to agriculture falters and ends?

Sustainable food production means **growing food consistently well within ecological limits and not constrained by the need for nutrient and energy inputs, especially when those inputs come from fossil fuels.** Think about this for a moment.

Australia is famous for its sheep. Even today, more than 70 million sheep are grazing across the dry continent, producing wool and meat. In 2021, this national flock converted grass into 345,000 tonnes of wool that contained roughly 86,250 tonnes of nitrogen. In 2023, the national flock also delivered 740,000 tonnes of meat, over half of which was exported. That is another 74,000 tonnes of nitrogen that left the paddocks.

Removing that much of a critical nutrient from the soil is not sustainable. Mining is only sustainable if there is an infinite resource being extracted. Nitrogen in soil is not infinite, so the only way for production to continue is if the farmers apply fertilisers to replace the nitrogen. The same logic applies to all the other nutrients and the soil carbon.

Sustainable food is not a type but a reflection of the production system. Indeed, there are nearly as many production systems as there are producers. There will always be any number of production methods, and many can be sustainable with efficient use of ecological principles, what we call food ecology. There is no one sustainable food, no one size to fit all. In short, sustainable food is all about food ecology and context.

Sustainable food **is understood scientifically, is flexible, and works with the constraints of the producer and the production environment.** Sustainable food is also about

- what kind of food is produced—the nutrient density and

calorific content determine the quality of the collective diet. This is as much about what is grown as it is about demand.

- accessibility—there is no point in producing food at $10 a kilo if half the population cannot afford to buy it. Good food must, therefore, meet demand equitably. This is about price but also about availability and distribution.
- the 80:20 rule to benefit the 20% of producers who are getting it right and delivering 80% of the food supply. It should also get some of the 80% of weaker producers up to speed.
- variety is important because humans need variety for a sustainable diet, but also because variety can deliver redundancy and resilience in production and food supply.

The size of the sustainable food challenge

Achieving food security for individuals and sustainable food for society is a vast undertaking. So far, humanity has delivered a miracle to keep up with our own numerical and economic success. In a classic chicken and egg contest, we have met a staggering increase in demand for food through intensification and the industrialisation of complex supply chains run on fossil energy. Only this is just the beginning.

We already know that future food security will mean at least 22 trillion kilocalories a day for another 100 years—sustaining this level of food production will be another miraculous achievement. Only we don't have 100 years' worth of oil and agricultural inputs, and business as usual is mining the soil, leaving it degraded in more than a third of fields and pastures. Production systems will have to change.

It will require an ecological focus even in the intensive systems of developed and emerging economies. And as mindful sceptics will already know, there is no single solution to this complex challenge.

How long will it take?

Earlier, I asserted that sustainable food is essential for feeding everyone well for as long as it takes. How long is that?

The global population of 8 billion people is still increasing at 8,000 per hour. By tomorrow, the number of people on the planet will have risen by 192,000 and by 1.34 million extra people by next week. In a year hence, barring a major global catastrophe, there will be 70 million extra people, nearly double the number of Canadians.

Numerically, the challenge gets increasingly wicked. The current paradigm is that we grow our way out of the problem. Global economic growth is a prickly subject, but everyone projects growth. This means more people with money to buy more energy and nutrient-dense food—the staggering growth in pork consumption in China is a classic example. Demographers tell us there is a transition that the human population could pass through to lower overall numbers as this extra wealth slows reproductive rates down to the replacement rate of a little over two children per woman—the best explanation of this transition comes from the late, great Hans Rosling. But this transition to the demographic dividend will take at least 100 years, perhaps longer. And as I explain in the *Mindful Sceptic Guide to Escaping the Malthusian Trap*, demography is complex. Meantime there are people to feed.

It is sobering to remember that the fossil fuel pulse will not last this long. The challenge could be cut drastically by collapse. But then, by definition, sustainable food will have failed.

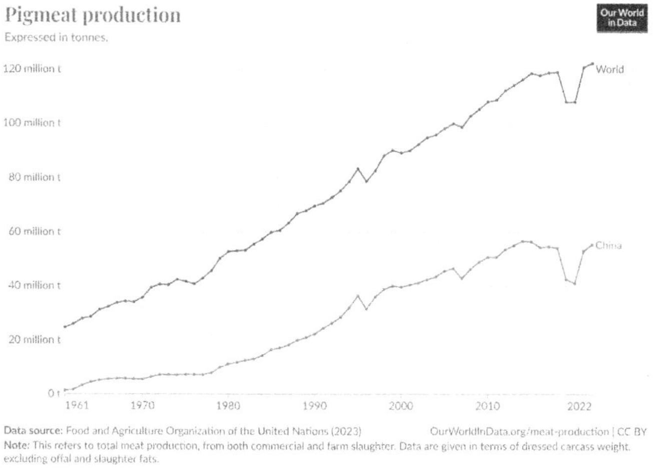

Sustainable food production requires concerted effort, radical change, and evidence. Food production is enough right now, but it is not sustainable. Knowledge of the supply and demand side of food must be spread widely, especially the role of the ecological engine that allows farmers to grow food.

Science needs to tell us how the ecological engine works. What fuel does it need? What is its optimal and repeatable running temperature? And when will the engine reach its local and global limits?

Knowledge can be distilled to provide producers with the evidence to make smarter production decisions from the smallholding to the large commercial properties. It can help investors with their returns and help policymakers design rules to help feed everyone.

Food production must happen for a long time to come. So, we also need to know what happened in the past to predict what future food production will look like. Current vested interests and a profit-driven production system will not take us beyond 2050; the trajectories are in the wrong direction. New financial instruments that prioritise long-term production over short-term profits are just one of the many options.

No other organism has appropriated nature to feed itself as humans now do. We have converted the landscape for agriculture, diverted the freshwater, mined the earth, fished the oceans, reduced global biodiversity, and even messed with the climate. At least a third of historical and current greenhouse gas emissions result from agricultural production. These global changes happened to **feed people who made more people.**

Today, this numerical, technological and economic success presents the biggest challenge humans have ever had to face. How to feed everyone well.

Fail, and humanity will collapse.

Key Points

- Agriculture enabled human settlements and communities, but the Industrial Revolution and the use of fossil fuels transformed agriculture into an energy-intensive industry, significantly boosting food production.

- Unprecedented global population growth from energy-hungry agriculture presents a colossal challenge in providing enough food for everyone.

- Sustainable food production means growing food within ecological limits and reducing dependency on fossil fuel inputs.

- Billions of people experience hunger and malnutrition despite sufficient food production because of a failure of one or more food security components: availability, access, utilisation, and stability.

- Ecological principles are essential to achieve sustainable food production and to secure a consistent food supply.

- Concerted efforts, evidence-based decision-making, and innovations are needed to ensure that food production can meet the needs of a growing global population while safeguarding the environment.

In the next chapter

Most of the food produced to feed everyone is grown in soil. How sustainable is the land management that makes it happen?

Chapter Two

Sustainable Land Management

Feeding everyone well requires sustainable land management because food still comes from the soil. Are we asking the right questions about it?

First, an oxymoron

An oxymoron is where contradictory terms appear in conjunction. How about this one? Sustainable land management.

In the professional community that I come from, naming sustainable land management an oxymoron is heresy. I am likely to be struck off. Many of my colleagues in applied ecology and natural resource management are in pursuit of SLM, to use its affectionate moniker. It is the cup with miraculous healing powers, providing eternal youth and sustenance in infinite abundance. Drink from the cup, and the lost art of sustainable agriculture and use of the land will come to you and make everything better.

The prophets of agroecology, regenerative agriculture and even the few eccentrics who want to pay back the soil carbon debt to fix climate change are firm advocates of sustainable land management and believe they already have the grail. They will flame me, too. If I leave myself open to ridicule, why be so contrary? First, we need to know what it is we are calling contradictory.

The FAO defines sustainable land management as

> *measures and practices adapted to biophysical and socio-economic conditions aimed at the protection, conservation and sustainable use of resources (soil, water and biodiversity) and the restoration of degraded natural resources and their ecosystem functions.*

Read that definition again, and you'll see where I get the oxymoron. It is impossible to manage land to use natural resources—we call most of this activity agriculture—and, at the same time, protect, conserve and restore. It is like trying to fill a bucket with a huge hole in it. Agriculture takes nutrients off the land and puts them into humans and the animals that humans eat. These nutrients are rarely, if ever, recaptured. Instead, they end up in oceans and rivers, sometimes via sewage treatment works. Over time this export results in the depletion of nutrients, soil carbon and soil biodiversity that lowers fertility and plant production. Farmers know this and replace the offtake with fertiliser inputs or move to new land when the soil is depleted. Agriculture has been like this everywhere since it was invented 12,000 years ago, but it has become a caricature of nutrient exploitation and soil degradation with intensive, industrial-scale agriculture only in the last 120 years.

Maybe I am being too harsh. It is possible to manage land with the future in mind or with an intent to maintain the resource rather than exploiting it, or both. But when has humanity ever been so thoughtful? Here is the thing.

Humans are the most successful species the planet has produced. This is true if success is measured in biomass, resource use, or the appropriation of resources that restrict availability to all the other species on the planet. It also applies to ingenuity, technology use, and adaptability. No other large mammal lives everywhere or comes close to the ability of *Homo sapiens* to decouple from ecological constraints.

And there is one huge success that has eluded all other species. We are unique in our ability to capture and use exogenous energy with a once-in-history outcome. We have **converted fossil fuels into 8 billion humans, adding 6.2 billion since 1900**—more on this later.

Such conspicuous success means we are forced to use natural resources to support the people we have made from the fossil fuel pulse. In the process, we are degrading natural resources, especially the soil, and are wholly dependent on the energy pulse that will come to an end. Despite all our successes, humanity is vulnerable to collapse and has failed to overcome our basic instinct to make more people—the global population is still increasing at 8,000 an hour.

We have no choice but to manage land for food production. Chaos will break out everywhere if we don't. Oxymoron it may be, but it is easy to see why sustainable land management is pursued with religious fervour through research and practice. If we don't manage land sustainably, then collapse from our dizzy heights is inevitable. The logic is sound, and the scientists have been looking into it for some time. However, I will pursue the oxymoron jibe by showing you a few of the research questions asked by seekers of sustainable land management truths.

Land managed primarily as vinyards and livestock production on the south coast of NSW, Australia.

Sustainable land management research questions

Science is all about asking and answering questions. A core skill of the researcher, and indeed the mindful sceptic, is to know what questions to ask. They must be relevant, add to the knowledge pile, and answerable through one or other version of the scientific method. Some researchers even follow a system for SMART objectives that are specific, measurable, achievable, relevant, and time-bound because an acronym for what you are trained to do seems to help. The kinds of questions researchers say are emerging for land management are grouped under three themes: 1) diversity of land systems, 2) place-based research and 3) new scenarios. I looked at one set of 'emerging questions' for sustainable land management from the scientific literature and present them verbatim with my commentary on their relevance to feeding everyone well, the ultimate objective of sustainable land management. In the spirit of the contrarian, this section is not comprehensive and will offend some researchers asking smarter questions. But the questions in italics below come from respected primary literature sources, so are taken seriously by the scientists.

I apologise in advance for the dry banality of these offerings and for the fact that they come from a different galaxy to the one Jane Doe exists in, but here goes.

Diversity of land systems

- *Are the correct drivers addressed for investigating solutions on sustainable land management, considering the knowledge, values and rules define the decision context?*

- *How to capture countries' activities and characteristics properly to account for emerging issues such as large scale land acquisition, or long-distance externalisation of effects within global agro-economic models?*

- *What are the options to govern land as a global commons?*

We already know the drivers. In no particular order, they are subsistence, profit, and expediency. This translates to land clearing, shifting agriculture, and a tendency toward intensification of production through energy and nutrient subsidies. It often leads to over-exploitation of the soil that introduces other drivers associated with soil degradation, such as yield loss, water stress, nutrient or soil carbon depletion, weeds, pests, and soil erosion... the list of immediate issues can be very long.

We also know that self-interest is powerful because farming is a high-input, high-risk operation that takes great courage. Few humans take on tasks with these characteristics without some reward, whether food for their family or profit.

We already track land use and have a good vision of land ownership, including deals to acquire land. Each jurisdiction has a known level of control over these outcomes. Modern models are already good at the externalities. Less good are the attempts to pull them in.

Why should we govern land as a global commons? Presumably, this is a reflex response to the historical exploitation of land. However, even if a communal model were more efficient, centuries of land rights and use are near impossible to reverse without a shift to authoritarian governance. Ironically, land rights may be a paradigm that needs reform because solutions are as much about land use as land management.

Place-based research

- *Which are the next steps to enable global agro-economic models to address a larger set of commodities, different land-holding systems, capture nutrient cycling and provide sufficient information on food security question on a finer spatial resolution?*
- *How can the tele-coupling concept be operationalized in research to better underpin and embed life cycle analysis in global relations?*
- *Which data gaps should be closed to better account for local variations in the socio-economic context of sustainable land management?*

Come again? Seriously, what on earth has this technobabble got to do with feeding everyone well?

Future perspectives and new scenarios

- *How to implement the mutual feedback of biodiversity and agri-*

cultural production in today's global model system estimation global agricultural yields and estimate optimum intensification levels?

- *To what extent do concepts like sustainable intensification that claim to have synergies between SDGs really have potential, what are the trade-offs hidden in these systems and in what local context are such concepts applicable?*

- *How can integrated scenarios capture the links between production and consumption, rebound effects and Jevons paradox?*

Coupling biodiversity with food production is a great idea. Figuring out how diversity and biological activity mesh, especially in soil, is helpful. It will generate critical insight into how much intensification is possible within the ecological integrity of a field or forest coop. Only we have something like this already. It's called **land managed within capability**, and it comes as a spatial data layer.

Intensification is another word for energy subsidy. Agriculture with inputs for tillage, fertiliser, irrigation, pesticides, and mechanical harvest adds external energy to the system's ecology. The trade-off for the extra production that inputs generate is that the system fails without them.

Intensification is a cover word for agriculture changing from a net energy source, with more food energy out than energy in, to a net energy sink. This is only possible because humans have acquired exogenous energy—energy from outside the body—from fossil fuels and had the ingenuity to build machines to apply the energy to fields, using more energy in the process.

Do we even know the links between production and consumption? They are many and complex. For example, why do we produce so many hyper-processed foods that are consumed to make people obese and yet undernourished? It's not rhetorical; the answer is profit.

Jevons paradox? Sounds like a bit of that snobbery to me. As I am sure Jane had no idea what it meant, I asked Google for a summary.

> *Jevons paradox occurs when technological progress or government policy increases the efficiency with which a resource is used, but the rate of consumption of that resource rises due to increasing demand. However, governments and environmentalists generally assume that efficiency gains will lower resource consumption, ignoring the possibility of the paradox arising*
>
> Wikipedia.

In short, never trust anything that comes out of economics.

Some sustainable land management experts excited by what a farmer has to tell them—he's the dude in the straw hat and split overalls.

Asking better sustainable land management questions.

Before I disappear into a hole of despair at my academic colleagues' naivety and intellectual snobbery, let's back up a minute. Taking a breath is good mindful sceptic practice, as is using the pause to expand our critical thinking. Let's ask a different question. What does the big picture tell us?

- There are 8 billion humans alive, 7% of humans that have ever lived, and our numbers are growing.
- Food production is adequate to feed everyone well for now, thanks to the fossil fuel pulse and an impressive six-continent food supply chain. Still, there are food security issues everywhere that stem from global and local inequality.
- Meanwhile, poor metabolic health is a pandemic, with a billion people overweight and 3 billion more suffering from one or other nutrient deficiency.
- Food still comes from the land, like it always has, and that land is degrading. The water to keep crops and livestock alive is depleted, often polluted and increasingly unreliable.
- The fossil fuel pulse will end.

The big picture suggests some equally big questions that bring us back towards sustainable land management.

- How should humans use the renewable and capital stocks of

planetary resources?

- Can we feed everyone well?
- And, if we can feed everyone, how do we do it without making a mess?
- What happens when the fossil inputs become too expensive for most farmers or become so scarce that farmers cannot access them?
- Should humane control of human population growth be implemented?
- Can our social contracts survive the resource squeeze?

I could go on, getting closer to what scientists might recognise as testable hypotheses. The truth is this: Humanity has to grow most of our food in the soil. And we need a lot of food, starting with 22 trillion kilocalories daily and enough nutrition to maintain metabolic health. Food insecurity already affects a billion people, and malnutrition affects at least half the people alive. Sustainable food is a big deal.

Food makes many of the big questions converge on what we do with land to grow and rear enough food to feed everyone well. And if we want to grow enough food without trashing other resources or failing to reverse biodiversity loss, then we have to grow that food ecologically. We have to get closer to production systems that reduce, reuse and recycle nutrients and carbon. We can even call this sustainable land management if we want, but the foundation is **food ecology**—the relationships among living organisms, including humans, and their physical environment for producing human food.

Tractable land management questions

Questions born of the big picture are not what the SLM researchers ask. When it comes to sustainable land management, we see the questions being asked are bonkers. "Blah, blah, blah", as Greta Thunberg so eloquently puts it. The overthinking pushes the actual questions and necessary pragmatism out of the way, and we are left with unanswerable drivel.

Instead of ranting further, here is what I suggest as more tractable and relevant sustainable land management questions that must be asked to set the context for the detailed hypotheses that follow:

- Who manages the land?
- Do they have the social licence to make land management decisions? More strictly, do we trust land managers to act sustainably?
- Do land managers know how to make good decisions? And do they have the capability and capacity to act on the knowledge?
- Can land managers resist the immediacy of their need for subsistence or profit? What are the social constraints on land management decisions?
- How long can soil health be maintained, given poor land management decisions? Can soil recover when rested?
- If nutrients are mined, what does it take to put them back without impacting soil function?

Develop these questions into research, and the answers are much more likely to help meet the feeding everyone well challenge, except that these questions lack political correctness. They are not just about people; they question their actions. Nobody will fund research that implies land managers make poor decisions. Food ecology avoids finger-pointing by focusing on the ecology of production.

Five principles of sustainable food and agriculture

Modern farming is all about simplification at scale. Production of a single crop or type of livestock is considered the most effective route to efficiency and profit. Couple a single product with large fields or big herds, and economies of scale add to profitability. Operations are easier if the producer knows about the thousand things to fix to grow a wheat crop and not have to worry about the thousand things to keep livestock alive. The prevalence of single-system farms across the agricultural landscapes of most mature economies suggests that this is true.

Take a look at this image of a wheat field.

Large field of wheat plants all the same age and size and nearly ready for harvest

The same plant, same size, planted at the same time, same maturity level and nearly ready for harvest at the same time across the horizon. The efficiency comes because there is a single objective: the production of a commodity for sale. Nature is harnessed to put all its energy and the vast energy subsidy provided by the farmer into one product, in this case, wheat.

A meadow used for animal production is also a managed system, but it looks very different. Numerous grasses, herbs, and forbs, along with invertebrate animals, might add up to 1,000 species or more that live permanently in a meadow. Superficially, the meadow looks similar to a wheat field. But look closely, and there is so much more going on.

Many species of plants and animals survive in this pasture

Ecology tells us that when an ecosystem contains a diversity of species, it is more efficient at cycling nutrients, is more productive over a wider range of conditions, and retains that productivity for longer. More diverse ecosystems are also more resistant to disturbance and bounce back after such shocks—they have resilience.

The reason for apparent stability is that nature is very good at using available resources. If some organic material is present, for example, a dead leaf, an organism will come along to exploit it. In a diverse system, resources are partitioned among many species, each slightly different in their use of the leaf. A consequence of this is a different kind of efficiency. Diverse systems recycle and conserve nutrients far better than simple ones. Efficient cycling from diversity is the fundamental premise and the logic behind the five principles outlined by the FAO for sustainable food systems. Here they are:

1. improving efficiency in the use of resources

2. conserving, protecting, and enhancing natural ecosystems

3. protecting and improving rural livelihoods, equity, and social well-being

4. enhancing the resilience of people, communities, and ecosystems, and

5. promoting good governance of both natural and human systems

The verbs here are instructive. There are verbs for caution—conserve, protect— and then there are development verbs—use, improve, enhance, promote. As with most UN directives, there is still tension between development and sustainability. So, would these five principles deliver sustainable food systems? With the diversity paradigm in mind and food ecology as our mindful sceptic foundation, let's closely examine each one.

FAO Principle #1—Improving efficiency in the use of resources.

Intensive agricultural production systems are efficient. They have high yields and generate huge calories in commodities that can be moved and processed in bulk. Intensive agriculture is profit-driven and minimises input costs throughout the production cycle, focusing on efficiency. The result is the global supply of cheap carbohydrate foods. The irony is that producers in this system are price takers due to global supply chain controls. They are always at the margins, which makes efficiency essential.

The sustainability problem is not that resource use is profligate—no farmer will throw fertiliser onto a paddock for fun—but that resource use is essential. Intensive agriculture is all about fossil fuel energy. Oil builds and drives the tractors. Oil and gas supply and fuel the fertiliser factories. Oil fuels the harvesters and the transportation of the product to processors. Intensive agriculture is energy-hungry. No amount of efficiency gains is enough to satiate such a greedy system. While efficiency gains are essential at all points in the supply chain, efficiency is not a sustainability concept if the production system uses resources.

FAO Principle #2—Conserving, protecting, and enhancing natural ecosystems.

Agriculture needs natural ecosystems as the source of critical ecosystem services. A mosaic production landscape interspersed with regenerating native vegetation is essential to maintain future productivity because a wholly artificial input system is not sustainable. Recall the monoculture of wheat in the previous section.

For example, controlling weeds, pests and diseases will be impossible in the global monocultures should any of the top five grains be impacted. Mosaics are the passive approach that at least offers a chance to repel invasives without compromising the efficient use of resources required by the first principle. The *Mindful Sceptic Guide to Biodiversity Loss* explores the idea that rewilding is a way to save the planet and that sharing, not sparing, is inevitable given the food demand projections from 8 billion souls. Although such a paradigm is challenging, it is possible despite many practical and social barriers to success. Conserving, protecting, and enhancing is possible, but a global response requires exceptional courage.

FAO Principle #3—Protecting and improving rural livelihoods, equity, and social well-being

Some form of social development is in the DNA of the United Nations and the FAO. The global reality is that inequity grows by the day. Aggregate metrics may have the global poor improving their livelihoods, but inequity grows as the rich continue to appropriate wealth via a nefarious financial system.

Food systems cannot be sustainable unless they deliver on these social metrics. Growing enough food is just the start. The right type of foods that are accessible to all is another matter. In the chase for cheap food that initially was a profit grab but is now a social necessity, we cannot increase food prices until people can afford them or risk the destabilising effects of hungry people. Note that this is not a problem hidden in the developing world. For example, before the COVID pandemic, two soup vans from the St Vincent de Paul charity handed out about 12,000 meals in inner city Melbourne, Australia, a thoroughly modern city of 5 million people with an average wage of nearly $200 per day. Then Covid hit, and demand exploded. In the 2020-21 financial year, Vinnies soup vans handed out 373,000 meals and now run nine separate vans.

Hunger is a growing problem, with the irony that obesity from excess cheap, processed foods is now just as acute. Cheap foods are not always healthy when they are the only foods you can afford or access.

FAO Principle #4—Enhancing the resilience of people, communities, and ecosystems

We always say that feeding everyone well is a people problem. Sure, there is plenty of ecology and nature, but people are using resources, managing land, and converting primary production into food and fibre. People eat the food and choose what food they eat or not.

The thing is that people are already really good at resilience. They routinely survive war, famine, and conquest. Even death is beaten in the aggregate—around 10% of the humans who have ever lived are alive in 2022. We are good at flexibility, surviving harsh conditions and living from hand to mouth if needed. Humans are extraordinary survivors. So much for horsemen.

Ecosystems are less resilient than humans. They cannot move to new regions (the great expansion), appropriate new energy sources (the subsidy from fossil fuel) or invent technologies to buffer against the weather (the Burj Khalifa in Dubai). Ecosystems respond to changing conditions through a turnover of species (ecological succession) where organisms unable to tolerate new requirements are replaced by another that can. Replacement implies source populations or rapid evolution of the species present. This has worked well for the entirety of life on Earth.

Disturbance unsettles an ecosystem, and it changes to accommodate the changes.

The problem for humans is that a changed ecosystem may not provide all the goods and ecosystem services we need or want. The ecosystem is resilient in its own time, but we need it to get back to help us out today. It is strange to lump humans and ecosystems under the same principle.

FAO Principle #5—Promoting good governance of both natural and human systems

This may be the only principle that matters in this shortlist of five. None of the others has any chance unless there is a paradigm shift in human attitude towards resources, the environment, and our place in it.

We are energetically subject to the laws of nature, just like all the other wild beasts. Our physiology generates a remarkable hold on entropy, so we persist long enough to reproduce, a defiance that uses energy and generates waste—a requirement of the Second Law of Thermodynamics. Obeying the law sets up a resource requirement for clean air, water, food and shelter that we cannot escape. But this is not the human belief system. Religions involving deities, our cultures and the religion of money all tell us humans are above nature, better than it, and separate at the pinnacle of evolution. As the bible tells us, we have dominion. There are now so many people that dominion has produced resource depletion and pollution of the planet. The risk of collapse is real and present. Our governance needs an overhaul.

57,000 people live in Greenland even though maximum temperatures are below freezing in winter and only reach 10C in the height of summer—these people cannot survive without energy subsidies

Ten laws for sustainable land management

Perhaps principles are not enough, and you prefer something stronger. Soil scientist Dr Rattan Lal, the winner of the 2020 World Food Prize for his work *developing and mainstreaming a soil-centric approach to increasing food production that restores and conserves natural resources and mitigates climate change* and international advocate for soil carbon, has proposed ten "laws" for developing sustainable land management practices.

Before we dive in, they are not laws. There are very few laws of nature, and it will take a very long time before scientists can declare them as such. Calling a set of principles 'laws' is clickbait, but we like the attention it brings to labeling them boldly. I should warn you that some of the proposals can get a bit technical but it's worth persisting because they bring us back to sustainable land management. Here we go.

#1 The biophysical process of soil degradation is driven by economic, social and political forces

The FAO tells us that 40% or more of global agricultural soil is degraded, a frightening statistic I explore in the *Mindful Sceptic Guide to Food, People and Soil*. By degraded, the FAO mean the soil is not producing plant biomass at or near its potential, given the climate and weather conditions. For degradation to happen, the natural flow of nutrients from the soil to plants and back again, a process mediated by the organisms and fuelled by carbon, is disrupted.

What Ratten Lal's law suggests is that disruption comes about through land management. Farmers, foresters and other land managers make decisions on production or what to do with vegetation that is influenced by money and society. I agree with this, and it is an important realisation. Soils are robust to the natural effects of weather events and climate change. Agricultural production systems—vegetation clearing, tillage, fertiliser additions, pesticide applications—are disturbances outside the normal range for intensity and frequency. By normal, I mean the environmental variability that occurs at a place over ecological time, typically centuries to millennia. These unusual pressures change soil structure, biology and carbon content and tend to reduce the growth of plants and below ground. In short, less biomass.

#2 The stewardship concept is only relevant when basic necessities are met

It is often said that the farmer cannot be green if he is in the red. The same applies to the consumer. Faced with high prices or food shortages, people do what it takes to feed their families. Expediency for such a fundamental need is a powerful force. I agree with this one, and it points to concepts of universal basic income or well-being guarantees provided under the social contract or at least the recognition that market systems do not guarantee food for everyone.

#3 The nutrient and carbon pools in soils can only be maintained if all outputs are balanced by the inputs

New South Wales, the Australian state where I live, is huge, covering some 801,150 km^2, roughly half of which is rangeland. Sheep are the most popular livestock, with 150 years of at least 40 million sheep grazing that native vegetation.

A fleece is roughly 10% nitrogen and for years farmers have sheared sheep and exported the wool to overseas markets. Fleece weight has increased over the years, but a conservative 5kg means 500g of nitrogen per fleece. Assuming 100% of the herd is shorn, and the minimum herd size is 40 million then 20,000 tons of nitrogen per year are exported from the soils just in wool or well over 3 million tonnes of nitrogen since 1880.

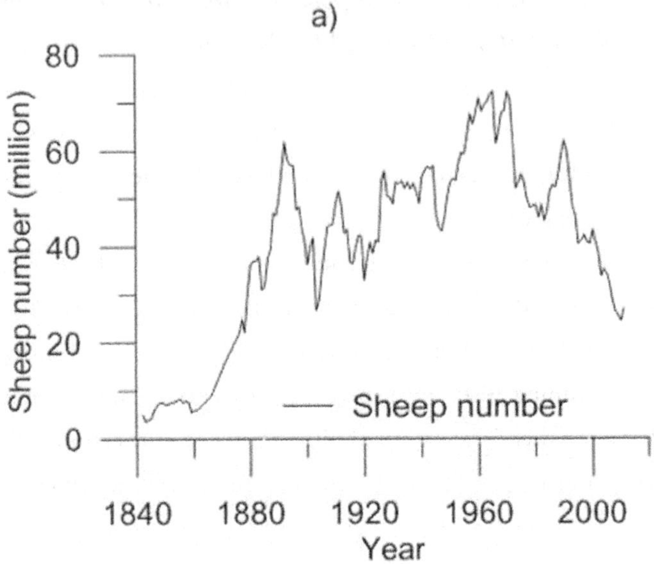

Source: Rančić, A. (2014). Reconstruction of Bore Hydrograph Trends in Fractured Rock Aquifers Using Data Mining Techniques (Doctoral dissertation, University of South Wales).

At this production scale, it is hard to balance this output level with inputs in both time and space. The reality was that it did not happen, and the grazing systems degraded. Agricultural production shows a similar pattern of net nutrient export from the soil everywhere you look. It is a robust law this one.

#4 Marginal soils cultivated with marginal inputs produce marginal yields and support marginal living.

There is only so much potential for soil to promote plant productivity. How much depends on the ability of the soil to retain moisture, provide a substrate for roots and a medium to exchange nutrients. Science has established what combinations of soil minerals, especially clays and organic matter, deliver efficient nutrient retention and exchange. This settled science means predicting production potential for each production system, and locality is possible. It is why crops are grown here and sheep reared over there.

Silk purses and sow ears apply to soils, too. Land use makes the general distinctions for this law. Arable agriculture takes the good soils, grazing land the poorer soils and forestry the dregs. Other land uses grab what they can, except for urban land and infrastructure, which always take priority. What is true is that a significant proportion of the 500 million farmers are operating at the margins.

#5 Plants cannot differentiate between nutrients supplied through organic or inorganic sources; it is a question of logistics, cost and availability

This is true and leads to the reality that when plants with high nutrient demand are grown—we call them crops—should the nutrient supply stop, so does the plant. Once a production system has mined the soil of nutrients and becomes reliant on nutrient inputs, this reliance is near total. Fail to supply inputs, and production fails. But the opposite holds, too. If a farmer increases ground cover, uses deep-rooted perennial plants and adds manure as fertiliser, plants respond just as well as if the crop sprayer had dumped liquid NPK. This is an important law.

#6 Agricultural soils can be a net source or sink for greenhouse gases depending on land use and management

I agree and can illustrate why with this graphic stylized from any number of studies that measure soil carbon before and after various land management interventions. The pattern is consistent everywhere. Although land management interventions do not always see an uptick in natural capital stocks or flows.

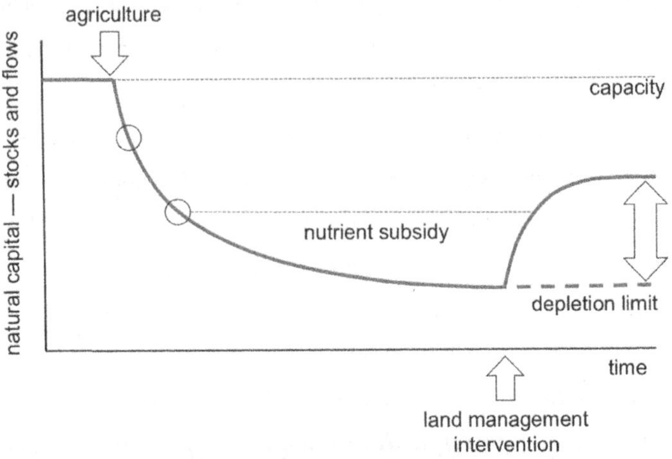

Source: Alloporus

Clearing native vegetation for agriculture releases CO_2, roughly a third of total human-generated greenhouse gas emissions. Soils used for agriculture tend to lose carbon over time because they lack vegetation cover or are depleted as crops and livestock yields leave the paddocks.

Management to reduce tillage, keep vegetation cover, maintain deep-rooted perennial plants and water management can promote soil biological activity and raise soil carbon levels, making soils a net sink. Crucially, source or sink is a choice. More on this in the *Mindful Sceptic Guide to The Soil Carbon Debt*.

#7 CO_2 released through the mineralisation of soil organic matter has the same global warming effect as CO_2 from burning fossil fuels.

Basic chemistry cannot lie.

#8 The yield potential of improved germplasm can be realised only if grown under optimal soils and agronomic conditions.

Intensification of agriculture is all about optimisation, the fine-tuning of production for maximum efficiency. Not optimum, maximum. Crop varieties bred and selected for the best possible yield do not perform if the weather changes or the soil conditions deteriorate. This introduces risk to yield levels that compound the risk taken to intensify in the first place. I like this law because it describes the reality of agricultural intensification and overlaps with laws #3, #4, and #5.

#9 Sustainable land management is the engine of economic development, political stability, and transformation of rural communities, particularly in developing countries.

Sustainable land management practices should apply to all agriculture all the time. COVID has shown how fragile food supply chains can be, and the wars in Europe and the Middle East will only worsen matters. Food security is rapidly becoming a problem everywhere and will be acute when the rich countries start having trouble buying their food supplies—yes, the UK, I am looking at you.

It would be good if the land value were again determined by its production potential. Well-managed land could then be traded at a premium. Unfortunately, demand will continue to push up land value irrespective of how much food can be grown.

This law bites hard when the 500 million farms that feed 3 billion people through subsistence agriculture can no longer use the shifting technique that allows depleted soils to recover under forest regeneration. Then they become input systems or crops fail.

#10 Sustainable management of soil implies the use of modern innovation built upon traditional knowledge

Yes and no. Yes, because innovation is essential to retain and restore nutrient and carbon capacity in soil. No, because whilst traditional methods are quaint, they are not always designed for long-term production, especially under changing environmental conditions or higher densities of people.

Yes, because tradition has tremendous inertia. It is easier to build on agricultural practices rather than replace them. No, because the assumption is that the traditional methods are worth building upon. This might be true to support a human population of fewer than 2 billion people—the number of people alive in 1900 before we started using oil in internal combustion engines. Most of these pre-industrial oil traditions were designed to feed relatively few people. Feeding a few might not be the best place to start when there are 8 billion people and their pets and livestock to feed—there are 6 billion more of us than when the first gasoline-powered tractors appeared.

One in five children younger than five are 'stunted' – they are significantly shorter than the average for their age, as a consequence of poor nutrition or repeated infection.

Humans are exceptional problem solvers. We are endlessly adaptive and create amazing innovations in a heartbeat. One of the advantages of so many people alive—some 10% of all humans that have ever lived—is the brainpower and ingenuity available to solve intractable problems.

Except people have issues we are not so good at resolving, at least in the aggregate. Technology in the hands of skilled humans can fix a faulty valve in an engine or even in a heart. Not so easy is to transcend the paradigm of dominion and resource use that we see as a right.

Intensification of food production systems through energy subsidies and technology has created near-endless cheap food that has become more and more people. Now that so many people are here, sustaining them is a priority. It is no surprise the FAO promotes sustainable food systems.

There is nothing wrong with good law. Even a good principle that is a guide rather than a rule that regulates actions with penalties for non-compliance can help achieve a collective good. Rattan Lal's ten laws for land management recognise the science but also the social reality that humans will meet their needs first. Not many people put the environment before their family or themselves, especially when they live on the margins. Humans will make resource-use choices based on immediate needs. The laws recognise the basic chemistry and physics of how soils produce food. Outputs need inputs, or the soil is mined for its nutrients and carbon. This is common sense that is often forgotten.

People create wealth from natural capital. There is no logical reason for this to be sustainable beyond the lifetime of the individual or perhaps a family lineage. Genuine sustainability over ecological and evolutionary time is not built into nature or human DNA. Sustainability must be an active choice, one that might go against intuition, tradition or immediate benefit. Applied to land management, this means that we might need real laws on statutes with strong enforcement rather than principles to achieve food security.

My take is that the principles outlined to achieve food sustainability are laudable but miss the point. Ratten Lal's 'laws' are better, but the truth is quite simple—it's a people problem.

Key Points

- Sustainable Land Management is aspirational, given the inherent conflict between resource use for agriculture and conservation goals. The historical depletion of nutrients, soil carbon, and soil biodiversity due to agriculture suggests that SLM faces considerable challenges reconciling these conflicting objectives.

- A global population of 8 billion people challenges sustainable food production with or without the preservation of natural resources. Agriculture relies on finite resources and fossil fuels, so food production is vulnerable to resource scarcity and environmental degradation.

- Research questions and the five principles outlined by the FAO for sustainable food systems may not align with resource-intensive agricultural practices, highlighting the tension between development and sustainability in these principles.

- Ten "laws" proposed by soil scientist Dr Rattan Lal for sustainable land management emphasize the role of economic, social, and political forces in soil degradation, the importance of balanced nutrient and carbon inputs and outputs, and the impact of land use on greenhouse gas emissions.

In the next chapter

Feeding everyone well and for a long time is a huge challenge and it's no surprise that sustainability is bandied around as an essential attribute of land management. But some basic ecology is required before we tackle growing the food that sustainable land management will help deliver.

Chapter Three

What is food ecology?

Food ecology is the engineroom of food production for subsistence and intensive agriculture. Here, we explain why understanding ecology is central to feeding everyone well.

Some basic ecology

Food ecology is how nature responds to the insatiable human need for food production and nutrition—the interactions between organisms that grow into food for humans and how humans gain nutritional health from the food we choose to eat.

Food ecology appears all along the six-continent global food supply chain in production, harvesting, preservation, storage, transport, processing, packaging, trade, distribution, preparation, composition, and consumption of food, as well as disposal of waste materials. It is also about human ecology, the complex interaction between the human body and the food we consume, that includes our evolutionary history and the health and nutrition of our food.

Here is ecology, the science that might help save us from ourselves.

Ecology has matured into a respected science discipline in its own right

What is ecology?

Ecology comes from the Greek word Oikos, which means house, home and family in ancient Greek. It is the science that studies the **relationships among living organisms, including humans, and their physical environment.**

> *Ecology (from Greek: οἶκος, "house" and -λογία, "study of") is a branch of biology concerning the spatial and temporal patterns of the distribution and abundance of organisms, including the causes and consequences.*
>
> Wikipedia

Ecology is about where an organism lives, how it performs within its home, and the effects of all the interactions with other organisms over its lifetime. All organisms have an ecology, and increasingly, it is manipulated by humans as we alter nature to support our own.

Science of ecology

Beginning in the Renaissance, 300 years before Linnaeus and long before ecology became a thing, scientists were content with defining nature through the fundamentals of physics, chemistry and biology. They searched for unifying laws of nature that could explain observable facts. And over time, many of the fundamentals were understood, providing the foundation for technology, engineering and agriculture.

In the natural world, biology helped us understand physiology, nutrition and how organisms maintained their chemical engines and from that, we developed medicine to treat our ailments. A historical perspective on nature and comparing the diversity of life—what we now call biodiversity—gave us evolutionary biology. This helped explore the origins of life and why nature is so diverse.

The word ecology is relatively new. It first appeared around 1700 with Carl Linnaeus but only developed into a scientific discipline in the 1950s. This makes it a science infant.

We needed ecology to explain what happens when organisms interact daily, for there are patterns and emergent properties not readily described or predicted by biology, chemistry and physics. Properties emerge when organisms interact; without ecology, there is little predictive ability to determine them. Some individuals survive while others do not. Some compete and win, others lose, and species occur together or do not.

- What happens to seedlings if ten seeds germinate in the same patch of soil compared to when only one seed germinates?

- How does the plant respond to the attention of a grasshopper or a goat?
- What does the goat do when the eagle flies above its kid?

Then there is an organism's response to the physicality of its 'home' or niche.

- What happens when it gets cold or hot or dry
- What does the organism do if a hurricane flattens half the trees in its habitat or a fire burns everything?
- What does the prairie dog do if a competitor steals its burrow or a rattlesnake finds the burrow entrance?

Ecology is the science that answers these and many more interaction questions. It can apply to the microscopic interactions in the tiny spaces between soil particles up to mass migrations of birds and grazing mammals across continents. All the various definitions of ecology revolve around the interactions between organisms and their environment, where the environment includes the other organisms. Ecology is about the consequences of interactions between organisms, including the transfer of energy and nutrients, making ecology also about processes.

- A plant uses solar energy and soil nutrients to survive, grow and reproduce—it makes more.
- A grasshopper, unable to produce energy, eats the plant leaves to power its survival, growth and reproduction.

- A stork eats the grasshopper because it can use the nutrients that the grasshopper accumulates from the plant for its survival, growth, and reproduction.

Energy is transferred across trophic levels from plant to herbivore to carnivore. Soon enough, the plant, grasshopper and stork die, and their tissues return to the earth as fuel for another combination of decomposer organisms that break the tissues into nutrients taken up by subsequent generations of plants. These ecological processes of energy capture and transfer power nature. They also power food production for humans. We are here, and understanding ecology is how humanity might persist. I think of food ecology through this systems lens as the processes that transfer nutrients and energy among and between organisms to generate biomass humans eat.

But we remember the past.

Evolutionary ecology—more making

Ecological processes come together to create more organisms (biomass) within the constraints of the environment and the interactions with other organisms. This happens through the behaviour of the organisms, their adaptations, and the limits dictated by the environmental context and conditions.

Organisms can only persist if they are already endowed with the form, physiology, and behaviour to survive, grow and reproduce in their environment. They have to fit. Evolutionary ecology is the science that helps us understand how this happens and how the fit survives and makes more. Here is evolutionary ecology in a nutshell.

- Organisms have attributes that are a product of their genes (genotype) and their interaction with the environment (phenotype)

- Change happens by chance mostly as disturbance.

- Organisms with the attributes to respond well to change do better than those without adequate adaptation and persist with a higher chance of reproduction—they are naturally selected.

- These naturally selected organisms do well for a time, reproducing more effectively than other types until they change across the generations through random gene mutation, sexual selection or until the environment changes.

- What was once optimal might not be anymore, and alternative types respond more efficiently to the changes and can be more effective at using resources.

- Organisms that responded well initially may die out in the new conditions.

- Some organisms are good at riding change by responding on the fly—a response called phenotypic plasticity—while others are stuck with their specific adaptations and must move to find conditions that suit them or rapidly evolve new ones.

- When acute change happens, organisms must adapt, move or die.

This sequence happens with disturbances large, small, acute or mild and has been on repeat for at least 3.7 billion years. All the organisms alive today, now called biodiversity, are the current expressions of this ancient process. In a collection of essays, G. Evelyn Hutchinson described *The Ecological Theater and the Evolutionary Play*, taking the analogy of ecology to be the many characters in a play about evolution. It is more like improv with cascades, feedback loops, and uncertain regulation features everywhere. Whatever the analogy, ecology is a science because there is more to nature than the sum of the parts.

A few essential features of this evolutionary process of natural selection that Charles Darwin and Alfred Russel Wallace figured out in the late 1850s are relevant to food ecology

- understanding ecology is impossible without knowing the fundamentals of evolution
- all organisms are a consequence of their evolutionary past
- disturbance is the process that maintains diversity
- extinction is inevitable

Homo sapiens are one product of evolution—a large mammal—with plastic responses to new environments, especially the food on offer. We became consummate explorers and colonists.

Then, we did something that no other large organisms could do effectively. We captured **exogenous energy**—we used energy sources outside our bodies to do some of our ecological work. It began with fire, extended to the domestication of animals and then went ballistic with coal, gas and, especially, oil. We used this energy to manipulate nature to work for us and change the environment to make us more comfortable. We built infrastructure and controlled ecological processes to grow food—we invented dominion. This gave humans energetic success, and we made more people, lots of them.

Applied food ecology

All human foods come from organisms—plants and animals—with an ecology. They have a 'home' that we modify to make them more productive but not always comfortable. They interact with each other and a myriad of other organisms, some of which we try to control. Most interactions are with organisms we cannot see and know very little about—the soil biodiversity and the microbial diversity that lives in and on us and our livestock.

Organisms that make up our food have a **food ecology** that determines what can be produced, where and when, and what food should be raised for the best nutrition and energy supply to an 8 billion-strong human population.

When humans invented agriculture 12,000 years ago, those early farmers refined their original process of obtaining food by gleaning what they could from nature's bounty to a process that manipulated nature. Agriculture altered ecological systems to simplify them and to channel energy and nutrients into crops and livestock. Essentially, they applied ecology.

Changes were modest at first, and then, with the help of fossil fuel energy and technology, became acute and widespread. A cornfield is far from the multi-species grassland grazed by bison and dug up by prairie dogs. But it couldn't happen everywhere. Roughly a third of the land area on earth is barren and uninhabitable. Forests and shrublands cover half the habitable area, leaving 48 million km^2 (46% of the habitable land) for agriculture. Only 23% of this land is used for crops (11 million km^2), yet humans appropriate over 40% of the global net primary production. Humans have applied ecology a lot.

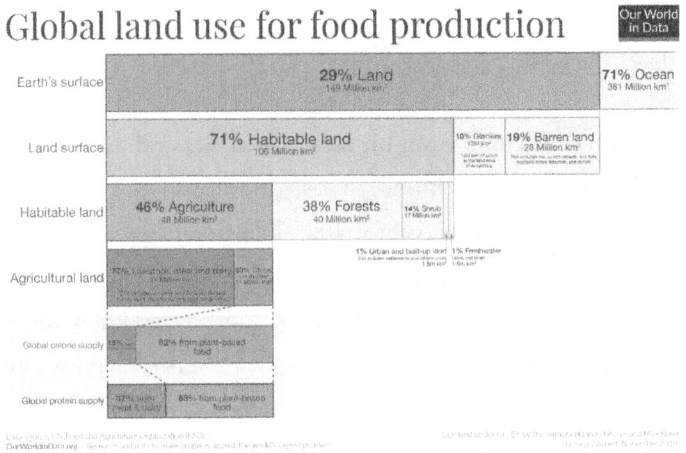

The challenge modern humans face is to feed ourselves. Growing enough food to feed everyone is a daily 22 trillion kilocalories mega task before we get to nutrition. We arrived at this sustainable food need by manipulating the ecology of half the habitable land and eating fossil fuels. The only way humanity will persist is ecologically—we cannot invent or power technologies to produce 22 trillion calories daily from thin air. That miracle can only happen via soil.

What food ecology is not.

The fundamentals of population, community, systems and evolutionary ecology, when used to understand human influence over nature, are called applied ecology. I distinguish food ecology because it focuses applied ecology onto the critical connection between food and diet and the answer to sustainable food.

As any mindful sceptic should, let's take a breath before we plough on. A brief pause to list what food ecology is not.

- **Not the environment**. It is what organisms do in environments and the consequences of their actions.

- **Not the conservation of nature.** This is an important topic that an understanding of ecology contributes to, but ecology is dynamic. It is all about processes and flows, not the specific organisms or their likely persistence—that is the domain of conservation and evolutionary biology.

- **Not the deep ecology movement**, the politics of returning to nature, or the growing dreadlocks. It is science and best understood through the scientific method.

- **Not scary.** It is complex and takes some effort to understand, but as the engine of food production, it is where the solutions to feeding everyone are found.

- **Not an opinion** but a science backed by an ever-increasing body of evidence. Much is known about how organisms interact to generate human values, and the evidence grows each year as the science of ecology matures.

What is food ecology?

If food ecology is not the environment, conservation, deep ecology and not an opinion, what is it? Food ecology is the relationships among living organisms, including humans, and their physical environment **for producing human food.** It is ecology applied to human food systems. And it is a vital science.

All ecology is dynamic and unstable in human timeframes, with any number of 'states' and rates of change. There is no 'good' state, and no one state is ever the same as the next, even in the same location. Ecology also differs from place to place.

One of the attractive features of ecology for scientists is this variability and complexity; we sometimes call it heterogeneity. It is also why we have to apply ecology to agricultural systems because there is a considerable risk in taking a complex, heterogeneous system and simplifying it to produce a single crop of one variety over a field of 5 ha. The vegetation preceding the field would have contained over a hundred plant species with thousands of individuals interacting. The soil that supported the plants contained uncounted species of micro-organisms and uncountable individuals, all interacting to transfer nutrients from organic matter to plant roots whilst helping to hold onto water. Even if the field, now ploughed with energy from oil, were left fallow, dozens of plant species would colonise the soil. Leave it like this for a decade, and a keen observer would see a change in the species as new ones arrive and the early colonisers die out. Ecology has happened. Ecology happens even when humans try to control, simplify, or direct it to produce what we value.

Food ecology is also about recognising that **sustainable food production is an ecological process, not an engineering one.** And to make it more challenging, most food ecology will happen in the soil where we can't see it and know little about what goes on.

The size of the food ecology challenge

In less than 100 years, the science of ecology has revealed a great deal about the transfer of nutrients and energy between organisms. Ecology has also developed to cover sub-disciplines to gain knowledge of organisms themselves (ecophysiology), their populations (population ecology) and how they come together in communities (community ecology).

While there is always more to discover, ecologists are well-informed about how nature works. Now, these learnings must be applied to the challenge of food production and diet. This is no small task when the global human population is 8 billion, and the food supply chains span six continents but still require vast inputs of exogenous energy and appropriate half the habitable land for agriculture.

To put this appropriation into context, 97% of the land mammal biomass on earth today—that's the weight of the mammals we can see with our eyes—is made up of humans and their livestock, animals that human beings manage for our benefit. You don't need to look far to find the real reason for biodiversity loss. Humans have used fossil fuel energy to simplify ecology on a global scale. I cover the details in the *Mindful Sceptic Guide to Biodiversity Loss.*

Initial success in agricultural intensification created more food that became more people, and now we are in a bind. We have made food production an industrial process that has to continue or people starve, only we have also tried to detach production from its ecology. That is both impossible and a colossal mistake. And if 22 trillion kilocalories a day for an indefinite number of generations defines the food ecology challenge, then we could throw resources at it and get going with the science.

But there is more...

Milk comes from the fridge

Ask a teenager where the milk comes from, and you know what she will say. Not able to detach herself from the screen in her hand, she is an individual, separate from the world and no more a part of ecology than the wildlife documentary she watched for five minutes.

And it's not her fault. She was raised in a society that believes in dominion over nature and that nature is there to be exploited as an inalienable right. We even avoid this truth in our environmental advocacy. Why are we obsessed with climate change when sustainable food is a far more immediate and potentially catastrophic challenge?

Although storm surges and wildfires are acute for an increasing number of people, climate warming and extreme weather make sustainable food harder to achieve, but it doesn't stop it altogether. Can you think what would happen if the food ran out?

We do everything we can to escape the admission that we are animals in nature. We even find it hard to see the ecology in and on our bodies—not least the billions of bacteria. Modern humans might not see ourselves as part of ecology, but we are organisms that rely on the ecological engine for survival. It might be easier to think that milk comes from the fridge, but the truth is that it comes from a modified ecological system, despite what our brain tells us.

The food ecology challenge is as much psychological as it is scientific.

Ultimately, the food ecology challenge is for people to embrace ecology and realise it can save humanity and the planet if we use its learnings wisely. This is blatantly humanistic on purpose because we are trying to understand how humanity can persist when there are 8 billion souls and counting. However, understanding and managing ecology well from the human-centric view to find sustainability will help stabilise human resource use. This will support other values of nature and leave some resources for other organisms, including the rare and endangered.

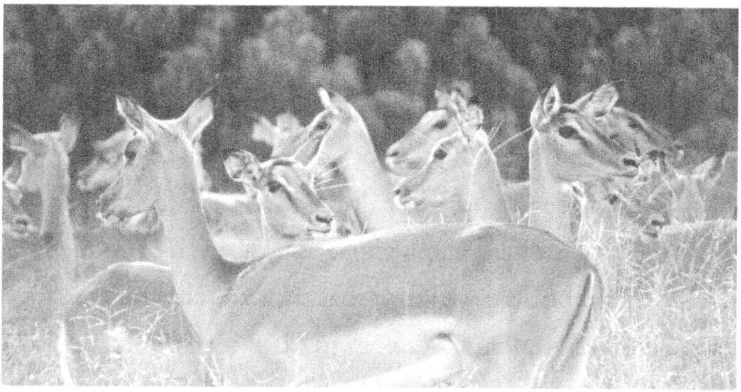

Impala are common grazing mammals in large parts of east and southern Africa. Photo by Alloporus

Our Mindful Sceptics concept of food ecology is to see it as the **engine of food production**—how organisms interact with the environment to generate biomass that humans direct into the food we need to feed everyone. This applied ecology is also about the supporting and foundation services for biomass production, including how to maintain and regenerate these processes where needed.

Food ecology is also about how food production can occur within ecological limits to not deplete the natural resource base, especially the soil. As I explain in the *Mindful Sceptic Guide to Food, People and Soil*, already 40% of global soils under agricultural production are degraded, with 70% of topsoil eroded. This cannot continue if humanity is to generate the necessary 22 trillion kilocalories per day for another 100 years at least to feed everyone.

Over-reliance on linear input-output systems of food production works for the moment but needs to be more sustainable for the duration. Half the nitrogen in human bodies comes from the Haber-Bosch process, the industrial production of ammonia that becomes nitrogen fertiliser. It's a crude metric, but half the food eaten contains nitrogen from man-made fertiliser. Regenerative and renewable production systems are essential because we know that the fossil fuel pulse will wane. This means understanding the ecology to adapt food production so it no longer mines the resource base but instead replenishes soil carbon and soil biodiversity.

Soon, agriculture must transition towards a net energy source rather than a sink with the plug out. The science of food ecology will help us stay on track and let us know if we have succeeded.

Key points

- Food ecology is the study of how nature responds to human needs for food production and nutrition. It focuses on understanding interactions between organisms, the entire food chain, and human dietary choices.

- Food ecology encompasses every aspect of the global food system, from production to disposal, as well as the intricate connection between the human body and the foods we consume.

- Understanding food ecology is crucial in human evolution and adaptation to changing diets.
- Humans are part of the ecological system, and ecological principles will deliver sustainable food production and the responsible management of natural resources.
- Regenerative and renewable food production systems are the means to address resource depletion and maintain ecological balance.

In the next chapter

Food ecology is a cute moniker for the science of food production, but what about the real world? What happens on the farm?

Chapter Four

Syntropic farming and the like

Sustained production and food delivery where and when needed, is a wicked challenge in need of audacious agricultural ideas.

Feeding Everyone Well

Syntropic farming appears in the *Mindful Sceptic Guide to Food, People and Soil*, highlighting its inventor, Swiss farmer Ernst Götsch, who developed techniques that reconcile agricultural production with landscape regeneration. Götsch's vision is to change how we perceive, interpret, and interact with the farm so that food production works harmoniously with nature, essentially food ecology.

More of the details later, but the idea is that food production comes about because of ecological processes—the interactions between organisms and their environment—that occur naturally. Each plant in a food production system grows in conditions that mimic nature. Farming creates an ecology similar to what would happen if nature were left to its own devices but with plant species humans find useful. Along with the other ecologically-based approaches to farming—agroecology, regenerative agriculture, holistic management, natural sequence farming, and rewilding—syntropic farming will be essential to feeding everyone well.

Feeding everyone well is not just about the fundamental human right of every person "*to an adequate standard of living for himself and his family, including adequate food, clothing and housing,*" established internationally since the 1948 Universal Declaration of Human Rights (UDHR). It is also about the survival of humanity. Society begins to unravel without adequate food, law and order break down, people move en masse to find what they need, and chaos readily ensues. It is this basic.

Predictions of a food supply catastrophe have little traction in mature economies. Abundant food on the shelves of supermarkets lulls consumers. Few have any idea that supply chains are fragile and production fickle. The good news is that preventing or at least reducing the risk of food shortage chaos requires three things: grow enough, distribute well and waste little.

One of the many ideas to **grow enough**, specifically, the how part of food production, is to embrace food ecology. A return to food grown more naturally within the constraints of the ecological context of the farm. Ecological farming comes in many guises, but first, we must understand intensive agriculture that mostly bypasses food ecology.

Modern agriculture

It sounds odd, but there are just two broad types of agriculture: subsistence and intensive.

Subsistence agriculture is generally small-scale, involves modest to no inputs and relies on soil's innate and regenerative capacity to deliver nutrients to crops and livestock. This method of farming is still the most common. The FAO estimates that five of every six farms in the world consist of less than two hectares, with more than 608 million small family farms across the globe that use around 80% of the world's farmland and produce 80% of the world's food in value terms. At least 2 billion people, one in four of the global population, in 500 million households survive as "smallholder" farmers, working less than 2 hectares (5 acres) of land.

Intensive agriculture is input-driven, typically large-scale and reliant on machinery. Inputs are energy, nutrients (fertiliser) and chemicals (pesticides and herbicides). These intensive production systems often involve irrigation or significant infrastructure such as glasshouses, feed-lots and rearing sheds. The energy subsidy to intensive agriculture is huge such that most intensive food production is an energy sink. This is a reversal of what made agriculture so successful in the beginning when it created an energy excess so that human societies could grow in number and support complex societies with priests, soldiers and rulers.

Most of the meat, dairy products, eggs, fruits, and vegetables available in supermarkets worldwide come from farms with intensive, high-productivity systems. Efficiency drives intensive production to create cheap food. Between 1930 and 2000, U.S. agricultural productivity rose by 2% annually, while the proportion of U.S. disposable income spent on food prepared at home decreased from 22% in 1950 to 7% by 2000. Intensive agriculture feeds close to 6 billion people every day on food that seems to get progressively cheaper.

The problem with having just two broad types of agriculture is the risk—a failure in either is catastrophic. Alternatives at this coarse scale add diversity to production and reduce the risk of failure. Options that cover a range of systems, activities and philosophies closer to nature exist. Let's take a look at one of them.

What is syntropic farming?

Syntropic farming is food production that attempts to be independent of inputs and irrigation and delivers ecosystem services, especially soil formation, regulation of microclimate and promotion of the water cycle alongside food production. An ecological approach to food production.

The creator of syntropic farming, Ernst Gotsch, advocates changing how we see, interpret and relate to nature. In his vision, holes become nests, seeds become genes, weeding becomes harvesting, competition gives way to cooperation, and pests and diseases are seen as the *'agents from the department of optimization of life processes'*.

Syntropic farming allows farmers to replicate and accelerate the natural processes of ecological succession and stratification. The idea is to give each plant the ideal conditions for its development to mimic what would happen in nature left to its own devices but with plants that humans find useful.

A farmer who learns the local succession, plants companions and prunes with care whilst maintaining ground cover and a diversity of crops and livestock should expect higher productivity by accelerating the natural processes. The idea is for ecology to operate synergistically and optimally for the local conditions, a production system close to what nature would produce if there were no humans to interfere. The reason for this is that nature is super-efficient. Organisms all strive to reproduce and come together in combinations that natural selection, through competition, determines as the most efficient.

Succession in ecology is like its namesake television drama. Plants follow each other and compete for future space, water and nutrients. Who wins the succession from the incumbents is less about patriarchal choice and more about plants' ability to compete, which, in turn, is influenced by chance, especially when they arrive and where, along with the adaptive responses they possess.

In the syntropic farming system, the farmer recognises the local succession and guides it with species and individual plants useful to long-term production. The guiding hand of the patriarch. As succession proceeds, so does the complexity. More species in multiple layers with greater numbers of interactions. Complexity tends to add ecological redundancy where more than one species supply processes so that if a key plant is lost, another quickly takes over its functions. Logically, this leads to resilience in many ecological processes but also to the productivity of the farm. If the carrots get hit by a frost, there are always the turnips.

Some plants compete, but some are beneficial for each other. Planting a fruit tree next to a bean plant. Planting together as companion plants results in a higher density of plants that optimizes the use of space, sunlight, nutrient cycle, and mycorrhiza.

Modern arable agriculture tends to be a single crop based on the soil prepared for that plant species. Weeds are bad, and tillage (ploughing and raking) are considered essential so that for part of the cropping cycle, the soil is bare. Syntropic farming keeps soil covered with plants or organic matter (from weeding, pruning or removing plants) at all times. This helps to maintain soil moisture, increases infiltration, reduces erosion risk and, most importantly, helps to sequester carbon into the soil. Soil carbon is the fuel for soil biodiversity that, in turn, helps to mobilise plant nutrients.

Understanding syntropic farming means reconnecting to nature. As Ernst Gotsch continuously emphasises, this requires humans to observe, learn, and understand that '*the mistake was when humans thought they were the most intelligent species, but we are only part of an intelligent system.*' Pruning and cutting branches to rejuvenate maturing plants accelerate the growth rate, increases the amount of sunlight for other plants, increases the nutrient cycle, and directs the process of natural succession. This is the trickiest part of this system. The farmer needs to look and listen to know when a plant will benefit from pruning as a stimulant and where it mimics the natural grazing and browsing by herbivores in natural ecosystems.

Farmers are not good with mistakes. An error or misstep in production can ruin a crop. This is not just a financial risk it can be evident to the neighbours. Farmers hate admitting that they 'stuffed up' even though they are human like the rest of us. Mistakes happen.

Syntropic farming, like several alternative production systems, uses mistakes. Looking, listening and learning the pattern of succession benefits from mistakes, so long as the error is recognised and the consequences understood. Ernst Gotsch takes this further and advocates for trial and error to introduce some planned mistakes and careful observation of the consequences as a way to learn more about the farm's ecology. Some basic knowledge of the scientific method would make trial and error super helpful.

Why service crops are an intelligent investment

Before the arrival of big agriculture with its large tractors, complex machinery, input heavy production of mostly monoculture crops on fields the size of suburbs, farming was a family affair. Everyone pitched in to help grow a dozen or more crops, rear livestock who ate most of the scraps and make home goods from jams and preserves to cheese and cured ham. Such variety required crops grown in patches or sequences and livestock to eat, provide draught power, and help fertilise the fields.

Some of the crops were never intended to be harvested. They were cover crops, green manures, trap crops, green bridges, and a variety of other multipurpose species designed to keep the soil healthy, reduce pests and weeds, and feed some of the animals who, in turn, fed nutrients back to the soil in the fields with productive crops.

Farming system researchers call these **service crops** because they provide regulation and support ecosystem services, which are important locally, regionally and even globally. Before production became an energy-intensive input system, non-harvested service crops were as important a tool as the plough. Then, almost overnight, they went out of fashion.

Agricultural census data in the US show that service crops are adopted in less than 5% of the total agricultural land. Data from Europe indicate that cover crops, plant residues and multi-annual plants occupy shares of 8, 7, and 8%, respectively in arable land during winter. However, the benefits of service crops have not gone away. Indeed, the agricultural research community has stacked up evidence for the good they do to yields and production efficiency and why this happens—in short, carbon gains, nitrogen fixation and more resilient soil biodiversity.

Intuitively, farmers know this, too, so why don't they plant them?

Knowing the old ways

The average age of farmers in Australia is 53 years compared to the median age of 40 for the rest of the workforce, whilst in the US and UK, it is 59. These folk are old enough to remember conversations with the blokes who farmed the old way before the fields got big and the tractors bigger.

Today's farmers know the value of cover on the soil and choose not to plant them. They claim that service crops are costly in time and resources, especially the limited availability (and high price) of seeds of several species used as service crops in local markets. In the cut and thrust of tight forward contracts and ever-higher input costs, the benefits to soil health are outweighed by the cash flow and resource use consequences of growing a crop with no immediate return.

The few service crops remaining are grown for fodder production. These crop species are suitable for grazing, but the plant architecture, growth habit, phenology or functional traits do not consistently deliver the ecosystem service benefits that the old-time farmers knew about without knowing the fancy jargon. The old stagers saw the benefits of deep roots, high leaf-to-stem ratios (or the opposite according to needs), symbiotic N fixation, drought resistance or high water use efficiency, short or long cycles, propensity for mycorrhizal infestations, attractiveness to pests and/or to natural enemies, competitive ability against weeds, allelopathy, and a host of other ecological traits without being able to describe any of them.

With few service crops, producers who have embraced intensification now rely on inputs to achieve what fallow land used to do—give the soil time to recover, aided by the soil biodiversity. The benefits of intensification are higher yields, short rotations that reduce the time between salable harvests, and volume. All good for a profit-orientated operation.

All is well for a while, perhaps a generation or two, until the soil is depleted and cannot hold onto the inputs or until the cost of those inputs rises higher than the returns.

Just add more land

The amount of agricultural land has increased exponentially since the start of the Industrial Revolution, mainly because chainsaws and tractors made it much easier to clear vegetation. It is easy to make the assumption that land can be added when more production is needed. However, according to UN data and independent research, the world has reached its peak agricultural land, and more land is being abandoned than is being brought into production. Roughly 5 billion hectares might be the peak.

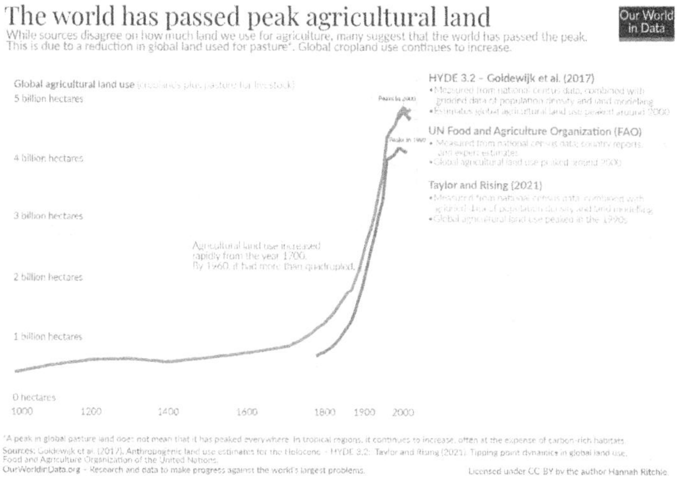

Not only is new land hard to find—not all land can be converted for crops or livestock—we have not noticed that some of the existing agricultural lands are already depleted and going out of production. This means it is foolish to assume that agricultural intensification is indefinite.

It all amounts to an egregious problem. So what can be done? Massive training programs in syntropic agriculture?

Audacious Agriculture Ideas for Australia

In an address to Congress on May 25, 1961, President John F. Kennedy set a national goal for the 1960s: '*landing a man on the Moon and returning him safely to the Earth*'. Ideas don't come much bigger, bolder and braver.

The Soviets had put a human-made object on the Moon 18 months earlier, presenting an opportunity for the idea. It took nearly a decade, but on Apollo 11, the United States put the first crewed mission on the Moon on 20 July 1969. Another five crewed landings followed between 1969 and 1972. In the decades since, no other country has managed the feat, and the United States is still the only country to have done it.

Being the first to achieve the impossible was a strong motivation for the Moon landings, as was beating the Soviets. Both helped galvanise a nation around a unique achievement. Putting a man on the moon is expensive and tested the commitment but it happened. Tragically, President Kennedy was assassinated on November 22, 1963, and did not see the idea come to fruition.

Motivation, near impossibility and commitment are just three requirements of audacious ideas. Growing enough food and getting it to everyone in nutritious combinations needs audacious agriculture ideas—big ideas, out-of-the-box suggestions, even ideas that would have no chance in the real world.

The 22 trillion challenge and the need to balance nutrition for metabolic health should motivate humanity to focus on modern agriculture as though it needed ideas equivalent to putting a man on the Moon. Feeding everyone well within planetary limits is nearly impossible but is more than an ideal or moral obligation because starvation is a powerful destabilising force. We need audacious agriculture ideas. Here are a few for our backyard in NSW, Australia.

Agriculture in NSW, Australia

I live in New South Wales, the largest of the eight states and territories that make up Australia. Agriculture is spread throughout the eastern two-thirds of NSW and covers more than 69 million ha of land or 50,000 km^2, more than twice the agricultural land in France.

Cattle, sheep and pigs are the predominant livestock since their importation during the earliest days of European settlement. About one-third of the country's sheep, one-fifth of its cattle, and one-third of pigs are produced in NSW. The state's farms also grow a large share of Australia's hay, fruit, legumes, lucerne, maize, nuts, wool, wheat, oats, oilseeds (about 51%), poultry, rice (about 99%), vegetables, fishing, including oyster farming, and forestry including wood chips. In 2019-20 the value of agricultural production in NSW approached AUD13 billion. This volume would put NSW 40th on the FAO country list for the value of agricultural production, with a higher value of agricultural commodities than 117 countries.

At this macro level, it all looks very promising. NSW has a diverse system of food production growing large volumes of food, more than enough to feed the local population with the excess sold to eager buyers overseas.

It looks less rosy for the farmer.

Merino sheep on a grazing property in NSW. Photo by Alloporus.

Drought, low soil fertility, pests, floods and wildfires are ongoing challenges to this food production that are risks to individual landholders. Add to this the financial stress from uncertain production, distance to local and international markets, high cost of inputs, shortages of seasonal labour, and being a farmer in NSW is a tough gig.

Many farmers have exhausted most farm-scale options to adapt to these challenges. They often need help to service high debt, have lowered their equity stake in the land, and are ageing rapidly with problems of farm succession everywhere. Bigger, audacious agriculture ideas could help to tackle the production problems. Here are three suggestions.

Resting the Western Division

The Western Division of NSW comprises 33 million ha of semi-arid land where sheep and goats scratch a life from degraded native vegetation. Stocking rates are incredibly low due to erratic rainfall, high temperatures in summer, and two centuries of grazing pressure from livestock and rabbits.

Any agricultural practice is a challenge in this region. So many sheep have grown on these rangelands that the soils are depleted from what was already a low nutrient base. The region needs a rest from livestock activity. Only the people cannot just walk away and leave the land for the weeds and feral animals.

There must be active management of the vegetation and pest species, but resting the land for a decade would dramatically improve future productivity. However, the local community cannot simply abandon the land and allow it to be overrun by weeds and wild animals. The audacious part of resting would be to subsidise somehow the land management that would need to continue and to keep people on the land to benefit when it is brought back into production. It will also mean wrestling with some legislation around leasehold obligation to graze the land. It is essential to manage the vegetation actively, but giving the land a ten-year rest period would significantly enhance its future productivity.

A hundred for a hundred

Ground cover is critical to production systems anywhere in the world, particularly in semi-arid areas or periods of drought when the soil can be exposed, dry out and end up elsewhere due to wind erosion. When the rains return, the water runs off the dry soil, increasing the risk of floods as the water fails to seep into the ground. Another audacious agriculture idea is for all agricultural properties in New South Wales to go for 100% ground cover 100% of the time.

During prolonged drought, 100% is impossible, so it is a stretch target, something to reach for rather than achieve all the time. The benefits of aiming for such a stretch will placate any disappointment. Perhaps incentives for cover that is readily monitored from satellite images and, dare I say it, penalties for non-compliance.

High-speed transport

One of the challenges farmers worldwide face is access to markets. Crops and livestock products quickly spoil if they are not stored, transported and processed in a timely fashion. Indeed, the prevalence of grain crops is as much to do with the robustness of the grain moving through the food supply chain as it is to the ease of production.

More than 85 per cent of Australians live within 50 kilometres of the coast. And it's not just surfing. A sea breeze on summer afternoons is a godsend. Domestic markets for agricultural products are concentrated in a few big cities and towns, making it hard to deliver perishable commodities from distant semi-circles around them.

If there were rapid transport systems such as high-speed rail from rural hubs into the cities, this would make it possible to grow fruit and vegetables at a distance.

Ecologically, all of the advantages of syntropic farming make sense and are consistent with the settled science. Carbon is essential for soil health, and moisture retention supports soil biodiversity. Together, they power the transfer of nutrients to plant roots. Multi-species systems are more resilient and frequently more productive than monocultures, especially when manipulated to be structurally complex and biodiverse.

However, a human is essential to syntropic farming and to most of the audacious ideas. The farmer has to pay attention, become flexible and be attuned to the natural dynamics of the ecology in his soil and vegetation. Typically it also takes more hands-on work with fewer energy subsidies and no big machines to till and harvest vast fields because the farm fields are smaller and interspersed with trees. This begs a critical question.

Can enough farmers pay attention?

They can and will because the alternative is going out of business, and very few farmers want to admit to that. Some farmers see service crops as an investment. They know that over time, it reduces external input costs, increases productivity, and provides long-term resilience. They put effort and funds into cover crops, legume leys, fodder crops, and green manures. Some also know why this happens.

Service crops are an opportunity to return some carbon to the soil. Carbon is the essential energy source, structural support, and matrix for nutrient and water exchange in soil, and a net increase toward the maximum local level supports productivity.

However, many farmers are stuck in a debt cycle or are accountable to the supply chain with farm gate contracts that make it hard to take any risks, even if it means adding more risk in the future. We have to help. Consumers can educate themselves on agriculture and realise the need to bring food production closer to nature. Not because it is green, lefty, or morally righteous, but because if we stay with an intensive production system with costly inputs, collapse is inevitable.

Audacious agriculture ideas are wild, out there with the fairies. The chance of them ever getting through the vetting process of politics or the famous pub test—the test of Australian public opinion—are slim, and this is as it should be. Just because an idea is audacious does not make it a good idea. Even the good ones may stumble over feasibility, practicality or financial constraints. The point is that we need to air them.

Key points

- Syntropic farming is an ecological approach to food production that aims to mimic natural ecological processes and harmonise food production with nature and promote biodiversity within farming systems.

- While subsistence farming relies on natural soil processes and is often small-scale, intensive agriculture is input-driven, large-scale, and heavily reliant on machinery, chemicals, and energy. Intensive agriculture has environmental impacts, consumes energy, and is vulnerable to supply chain disruptions.

- Ecological farming systems, including syntropic farming, agroecology, regenerative agriculture, holistic management, natural sequence farming, and rewilding, can reduce the risk of failure in agriculture.

- Modern farmers are challenged by climate change, soil degradation, financial stress, and market pressures. Innovative, audacious agricultural ideas are needed to address these challenges.

In the next chapter

The picture is big, and the only way to take it all in is to sample a few details. This is what we will do next, and in no particular order.

Chapter Five

A possum problem, among others

Feeding everyone well is a monumental challenge we have chosen to meet with intensive agriculture, and a six-content supply chain optimised for profit and efficiency. It's all good so long as the many risks are understood.

A possum problem

Where I live, there are possums. Cute little things with regular foraging routes. Each night they smell their way around our garden and, in the mating season, take frisky runs across the metal roof of our bungalow.

Our sneaky rutters are the common brushtail possum, *Trichosurus vulpecula*, a nocturnal, semi-arboreal marsupial of the taxonomic Family Phalangeridae, native to Australia, and the second-largest of the possums. They are a folivore that mainly eats leaves but have been known to eat fruits, small mammals and to lick the barbecue clean if you leave the lid open by mistake. Around human habitation, where the little darlings are at home, they are inventive and determined foragers who like fruit trees, vegetable gardens, and kitchen raids.

Our brushtails have a particularly delicate taste, and they enjoy anything I try to grow, especially leafy vegetables or herbs or the shoots of root vegetables. These items are manna from heaven. Native plants on the eastern seaboard of Australia are designed for hot and dry conditions. They have either spiny, narrow leaves or a waxy coat to retain moisture on hot summer days. The fire-tolerant eucalyptus almost encourages the fire to burn through the turpentines and phenolic oils in their leaves and twigs. So if you present a native herbivore and general forager with a delicate, bright green lettuce leaf, they love it. In the same way, the pet rabbit prefers kale over grass and hay.

Gardening in a possum habitat is a bit like putting an array of tasty treats in front of the kids; they will certainly not eat their greens if there are lollies and chocolate cake on the buffet. We don't have the space for the production volume to allow the sacrifice of a few leaves for the benefit of the possums whilst maintaining some sort of crop. So, the scale at which a gardener you produce food in this system struggles against the activities of the native animals.

People get around it by putting nets over their fruit trees and large cloches over the herb garden. A gent down the road from us has built walk-in 'greenhouses' made of mesh rather than glass. I watched him put up a new one recently. And at a rough estimate, I'd say he spent at least a thousand dollars on timber, meshing, various fittings, and fixings, not to mention the time it took him with his power tools. It seems unlikely that he would get a financial return from the crops he could grow in his $100m^2$ 'greenhouse'. Presumably, he enjoys growing the food and having homegrown vegetables with his steak. He was doing it for himself.

This anecdote explains why the modern Westerner has little chance of feeding himself if he escapes to the countryside and why losing plants to native herbivores is one of the significant challenges of integrating our production systems with nature. It also explains the dominion over nature paradigm because we have to change nature for it to perform to our advantage. Hence the large fields that overwhelm the occasional possum. Pesticides and herbicides deal with the organisms that can increase quickly when a food source is supplied. So, we have created systems that ensure a profitable yield is delivered.

The gentleman with the thousand-dollar-plus greenhouse takes his only option to produce homegrown food for his table. Before the construction effort, most likely, he'd see a few plants nibbled around the edges. Still, the crop is growing well, and he will be all excited until one night, the possums decide it is time for some lettuce, or a wallaby might hop up from the valley floor to avoid the heavy rainfall and take equal delight in the shallots and eat them all.

Integration is not easy

All this is to say that it has never been easy growing food. Placing our food production into nature involves significant compromise, especially in the first instance. But possums, wallabies, and bush rats can survive quite comfortably in the environment; they don't need the lettuces. If we integrate the systems well enough and ensure that it's only at the edges that these animals cause impacts, we can learn to live with them.

It's the quality of the native vegetation that we have to work on and ensure that we don't let our vegetation become too weedy. The natural vegetation must be looked after as much as the crops in the fields.

Allowing balance in the ecology of natural habitats as we try to inject ecology into our production systems will reduce loss and improve our ability to grow foods in a more ecological and sustainable way. There is a caveat to all of this. We can hear the farmers chirping at us, saying, "*Tell that to the rabbits and the pigs*".

In some situations, introduced organisms have adapted to foreign parts very effectively and become pests of even large-scale agriculture.

Rabbits are a classic example in Australia, where introducing a species from northern Europe not previously present on the continent but capable of reproducing rapidly on the grassy systems that exist in the continent became a dramatic pest of livestock production. Billions of rabbits have lived and died in Outback Australia. Controlling them was a challenging problem until the introduction of the calicivirus, and even now, as resistance emerges, rabbits are becoming a problem locally again and again. Similarly pigs. They are adaptable and intelligent animals that can utilise resources in the bush just as they've done in forests worldwide for millennia. Like the possums, they cannot resist the leafy greens of crops.

These are not trivial issues. Most pest control is reactive rather than proactive, with a limited understanding of how these animals impact the ecology. We can say the same about weeds, of course. Some weeds are native, but most are introduced and become a significant livestock production and cropping issue.

Control measures are not entirely technical. Most of the time, the solution has to have that pragmatic flavour, which recognises that it is as much about persuading people to do things differently as it is about letting ecology happen.

Fisticuffs in the aisles show supply chain fragility

Toilet tissue is a bulky item and takes up valuable real estate on the shelves of aisle 15 in the supermarket. Stock controllers work out the daily turnover and rely on the normal purchasing behaviour of customers to order enough to keep the shelves full just in time. Standard supply chain procedure optimises the number of deliveries to each store to avoid missing a sale.

Only during the COVID-19 lockdowns people panicked. They decided that toilet rolls were not just essential, but 'OMG, what would happen if we ran out?' The panickers bought double, three or four times the amount they would typically buy in a week and instantly, the shelves were empty.

It didn't take long in the first lockdown in Australia for toilet rolls to disappear from supermarket shelves. Although 4 billion people worldwide do not use toilet paper, Australians deem loo paper an essential service. They also forgot that there is sizeable local manufacturing, meaning normal supply was unlikely to be disrupted despite lockdowns. Instead, they bought up big. Terrible scenes of people hoarding five or six packets, even trolleys full of the stuff, with old folk going without and asking politely for just one pack—some individuals' selfishness was caught on video for all to see.

At the start of 2021, the UK was braced for a toilet roll situation, only this time with food supply and household consumables. Border changes from Brexit produced vast queues of lorries in and out of ports. According to the UK Department of Environment, Food and Rural Affairs, almost half the food eaten by the 60 million people in the UK comes from overseas. Others report that it might be 80% depending on whether it is tonnage or value, processed or raw.

Based on the farm-gate value of unprocessed food in 2019, half the UK supply was grown locally (55%), and the EU countries provided a quarter (26%), with Africa, Asia, and North and South America providing the rest. Remember that this is food for a population of 67 million people that need 146 billion kilocalories a day. The three largest value-added commodity groups were fruit & vegetables, meat and beverages.

Delays at borders can be absorbed if there is enough storage capacity to buffer any wait time, keeping food on supermarket shelves at the usual rate. There has been no need to invest in this capacity until now, and the market does not spend money on unnecessary services. In the absence of storage buffers, delays result in shortages. When the supply chain creaks, in this case, due to a political agenda, in the Sydney toilet roll debacle due to a health shutdown, there's a high risk of panic buying.

There are several realities here. Society relies heavily on supply chains. Almost all the people in mature economies shop for their food. Intrepid green-fingered types grow a bit here and there, but most people exchange money for food. Every person is reliant on a secure food supply.

We take this situation for granted. We assume governments work hard to ensure that food is readily available because they don't want it to fail on their watch. Failure is likely to get them ousted within an instant.

The reality is that supply chains are fragile, like wine glasses. It doesn't take much for a market-driven system to break. Each link in the food supply chain is optimised to maximise profit. Market forces define the system's efficiency within the constraints of individual efficiencies, available investment, and organisational capabilities. Markets optimise transfer (trades) rather than stability and resilience. For example, the level of turnover in warehouses is remarkable. They don't make money for retailers unless they are just temporary storage with goods moving in and out rapidly. No money is made if goods sit on warehouse shelves. Consequently, there's little storage capability and limited buffer should the system be interrupted.

The supply chain begins on the farm. The production end, how food is grown, is the natural focus of food security. Consumers can't eat unless the food is grown. It is essential to convert biomass potential into edible products, or the chain stops before it starts. But the supply chain itself is just as critical to food security. And how that supply chain delivers resilience rather than profit is where security will be gained or lost. Should the supermarket run out of bananas, what will the minions eat?

Most likely, COVID and Brexit issues are temporary problems. Politicians and the market will run around quickly to fix any market failure. And so the likelihood is that the food supply will be restored soon enough. In the meantime, there will still be something to eat on the shelves, even if it's packet noodles.

Starvation won't be the issue, at least not in mature economies. And in these statements is our risk. We assume that all will be well. But the level of disruption and the triggers from real or apparent shortages to basic instincts are huge. There will be fisticuffs in the aisles.

Perhaps we should grow more of our own food.

Why Chicken Coops Must Be Large

Neighbours of mine have chicken coops. They each have half a dozen hens who scuttle between the wooden shelter and a few square metres of grass constantly pecked. Like the wire greenhouse, these suburban coops of wood and wire cost way more to build than the value of the produce they house. My neighbours undoubtedly make delicious quiche, but I make my lunch much more cheaply from store-bought eggs. And they can't eat their hens if they want to eat tasty eggs.

When I was a kid, roast chicken was a Sunday treat. As an adult, chicken has become a staple. It is delicious, cheap and readily available in every form imaginable. We should venerate the humble chicken that can generate superfood and are themselves edible. In a way we do. Food and Agriculture Organization (FAO) statistics show a global chicken population of 34.4 billion chickens alive on earth on any given day, four for every person.

I was born in 1961, a fine George Clooney-level vintage. Growth in demand for chicken since the days my Mom would bundle the bird into the oven before galloping off to church has seen production grow from 6.58 billion chickens slaughtered in 1961 to 68.8 billion in 2018.

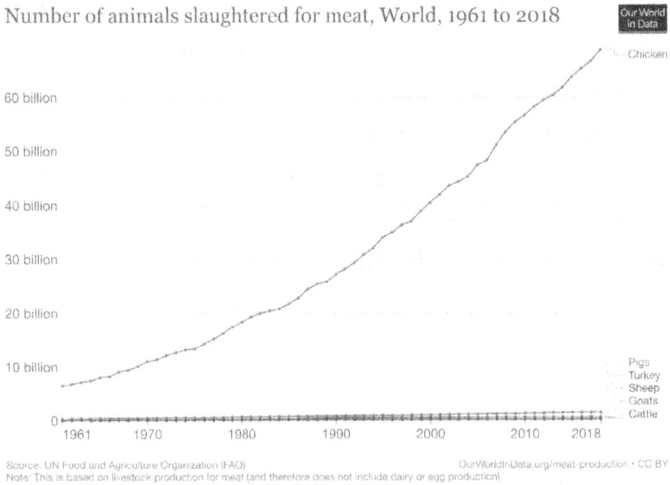

People like to eat chicken. And why not? Along with many tasty ways to prepare the meat, chicken is a lean, nutrient-dense protein. Nutritionists tell us that it should be on the menu. As more people can afford a healthy diet, demand will be there for the production curve to keep its upward trajectory. If the farmers can produce chickens at affordable prices, then there are people to buy them. Before getting carried away to the nearest KFC, let's see whether production can continue to grow.

Market for chickens from chicken coops

In a perfect market, processing should converge on the most efficient way to get a commodity to customers at affordable prices. Efficiency equates to profitability, and modern markets are a paragon of this powerful force. What happens is that technology and smart people with ideas come together to improve efficiency. Investors facilitate this interaction, and ideas become production systems.

When the market is food, technology has automated and homogenised production, processing and distribution to reduce the cost of each production unit. For chicken, the best returns have come by applying technology at scale with large chicken coops.

The executive officer of the Australian Chicken Growers Council, explains that over the last 20 years, smaller processors have been absorbed by larger ones, leading to decreasing levels of competition at the processing level and, eventually, monopsony conditions—a market structure in which a single buyer substantially controls the market as the major purchaser of goods and services.

In 2020, Inghams Enterprises and Baiada Poultry supplied 70% of Australia's chicken meat. And then 90% of the products from this raw material are supplied by just six processors. Such consolidation is the market rationalising efficiency.

Economies of scale also bring risk. Here is how one farmer sees the market for chicken in Australia...

> *"The biggest problem with the industry is the only way they can save money in growing chicken is by putting 100 sheds in one row – makes it a lot cheaper to transport out of that spot. "But what will happen is if we get avian influenza or Newcastle disease? It's going to wipe out not just three farms that are separated by kilometres like it usually does, it's going to take the whole operation out."*

Intensive farming is described as a high-input system. And it is. Fossil fuel energy in engines, fertilisers and infrastructure is added to the food ecology to supercharge production. But efficiency is also about economies of scale that drive consolidation and simplifying the market.

Obviously, as the number of players in the market declines, so does the level of competition. At the extreme, there is no competition and a single entity sets and takes the price. In the case of chicken coops, **this**

single entity is not the grower. And this presents a massive risk to the resilience of supply.

The endpoint of super efficiency

High demand is always a business opportunity that ignites powerful market forces. In theory, it brings competition to drive efficiency, capturing consumers with low prices for the commodities they want. The problem is the logical endpoint of super efficiency.

One or a few highly effective players corner the market by absorbing the less efficient competition, as in Australia with chicken production. This is good for the consumer until the efficiency confronts the risk of failure. Should the single producer with a handful of large chicken coops running on a similar production system fail, there is no cheap chicken.

We already know that efficiency gains have downsides when they do not cover the total production costs. It is always tempting to gain efficiency by accepting externalities through the supply chain and squeezing suppliers on price. Growers are forced into production decisions that are not in the animals' best interest or benefit the environment. Efficiency is the reason chicken coops have to be large.

We also know that exponential growth in producing anything is impossible when the resource it relies on is finite. Chickens can breed indefinitely only with sufficient food, water, space and lack of disease. Producers know this, but the market does not.

Humanity is locked into a dependency on efficient food production. Intensification continues to feed everyone well, especially those in urban areas without the means to grow their food. However, this powerful market force for efficient food production also increases production risk and reduces the resilience of the food supply. To be clear, the green line on the graph cannot rise forever.

Broad-scale land clearing is just lazy

In Australia, the term 'broad-scale land clearing' refers to the removal of native vegetation for development. Typically, this is to grow crops or create grassland that can rear livestock, but it also applies to land clearing for infrastructure, mining and settlement. Broad-scale land clearing makes agricultural land the matrix with patches of remnant native vegetation left behind.

Over the millennia, humans have preferred the small patches that require less energy to clear and maintain. Small areas are easier to manage and fit in the paradigm of moving on when the nutrients are depleted. We call this shifting agriculture. It is efficient and sustainable if there is enough recovery time before the next clearing event. It is also doable with available human labour supported by draft animals and simple tools.

Patches of vegetation altered for subsistence agriculture are not classified as broad-scale land clearing. Small patches create a mosaic landscape where the matrix remains native vegetation with gaps for crops and livestock.

Then, humans invented machines powered by oil. Chainsaws could cut down big trees in an instant. Ploughs attached to tractors could get bigger and easily pull across stony or heavy clay soils. It became possible to drive two tractors connected by a heavy chain across the land. The chain ripped the smaller trees and shrubs up from the roots.

Once the vegetation was felled, it could be gathered and burnt or left to decay over time. What was left was land that could be turned over and planted to whatever crop or pasture the farmer wanted. Efficient, sturdy and reliable machines did all the work for an initial investment and the price of the fuel. Using this capability to clear large areas to grow food at scale made sense.

Scale also made it possible to get more out of specialised machines for tillage, fertiliser application and harvesting. Soon, this newly intensive agriculture became an input system and a net energy sink. It was this intensification of agriculture that led to broad-scale land clearing. So why is it lazy?

Lazy Farmers?

It is a brave person who calls Australian farmers lazy. According to the government, crop farmers in this unforgiving land work an average of 52 hours per week, eight hours longer than workers elsewhere. They are also older than the broader workforce at an average of 52 years of age and retire later, if at all. Farmers have a deserved reputation as hard-working people who keep going long after the rest of us have stopped working and entered a grey nomad existence. Hopefully, the ivory towers offer protection from bold statements such as this one:

> *There are ways of growing beef that don't destroy the environment. Broad-scale land clearing is just lazy ... you don't*

have to clear habitats and drive species to extinction in order to get your food

Professor Brendan Wintle, Conservation Ecologist, University of Melbourne.

The implication here is that broad-scale land clearing is indiscriminate. It happens wherever land clearing is possible and where a crop can be grown or livestock reared. Farmers clear land whenever and wherever they can. In most countries, we believe that they don't because there are legal controls on where they can farm and how. And that farmers are smart enough to know where clearing will mean a crop and not a failure. But taken as a whole, farming has cleared or altered most of the native vegetation on over half the habitable land on the planet. Many conservationists argue for growing more food on the current agricultural land area.

The lazy accusation comes from the assumption that more effort on the existing agricultural land would produce equally viable yields. In short, farmers should be more thoughtful and not take the easy option of expanding production area at the expense of native vegetation.

The farmers will not recognise the nuance, but the agricultural system needs to be more active, not them. Intensive agriculture is driven by profit resulting from maximising production over effort. Get more with less, and the greater the profit from each unit of action or land, the better. Farming is often more efficient at scale, hence the need to clear large areas of operation. And there are benefits from newly cleared land.

What happens after broad-scale land clearing?

Cleared land begins its agricultural life full of nutrients, organic matter and potential. Plants tend to grow well so long as there is water and sunlight. Weeds can be a problem, as can pests on the newly homogenised land, but there is an innate store of nutrients and organic matter in the soil to promote plant growth. This production potential is why the land is cleared in the first place.

After a time, agricultural practices erode this capability. Nutrients are used up, and soil carbon declines when soil is exposed to the air. Often, production becomes marginal or unsustainable without inputs. Here is what this decline in natural capital looks like.

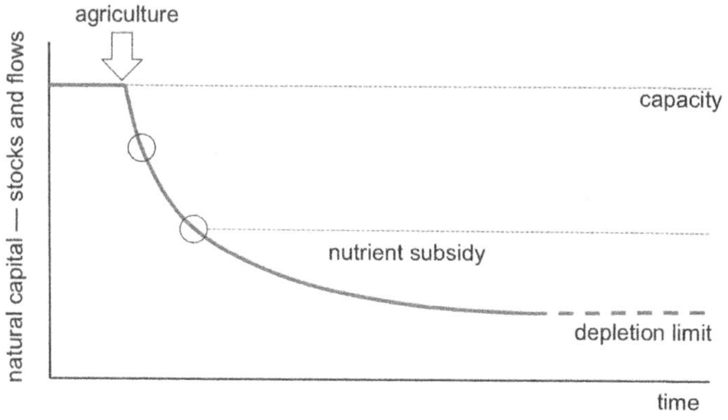

Once the land is marginal, three options are available to the farmer. 1) keep going as best he can to the depletion limit and hope that is still enough to eke out a living, 2) add inputs, especially fertilisers and water, to build crop yields back to profitable levels, or 3) move along to newly cleared land.

Option 3 is where the accusation of laziness appears. Rather than putting the effort, energy, and money into growing crops or animals on depleted natural capital, moving production onto freshly cleared land with better nutrients and water holding capacity is easier. Moving on is more profitable as it requires fewer inputs.

Conservationists prefer options 1 and 2, where farmers invest in existing agricultural land. They wish to prevent broad-scale land clearing for agriculture because, as I explain in the *Mindful Sceptic Guide to Biodiversity Loss*, clearing is the most significant cause of biodiversity loss and the source of a third of historical man-made greenhouse gas emissions. Options 1 and 2 cost the farmer. So the question becomes: who covers the cost?

The laziness of profit

Laziness here is not avoiding getting up at dawn and falling back through the farmhouse kitchen door 14 hours later. It is the laziness of profit. It is more profitable to clear new land and mine the natural capital of the soil than to throw inputs onto land that has been under production for many years. Even subsistence farmers know this as they practise shifting agriculture to avoid the problem.

Broad-scale land clearing is facilitated by machinery and energy but driven by profit. And anyone who has been in the world for more than five minutes knows that profit is a lazy bastard.

Resilience and resistance are not the same when growing food

Psychologists are keen on resilience, the capacity to recover from complex life events. A resilient person will adapt well in the face of adversity, trauma, tragedy, threats, or significant sources of stress—such as family and relationship issues, serious health challenges, or workplace and financial problems.

Resilience + resistance = stability

As you lie on the couch, no pun intended, the shrink will take you through four areas to help build your resilience—connection, wellness, healthy thinking, and meaning—each to gain the power to withstand and learn from complex and traumatic experiences. Mental and emotional resilience is crucial because plenty of life events can knock you down.

As an ecologist, I see the obvious parallels with how nature works. It is as though the highly trained health professionals borrowed resilience ideas from nature because nature appears stable. Plants grow, and animals survive as energy and nutrients move around in ecosystems that change but with bounds that make them look constant. Collapse is rare; production is the norm. What ecologists know is that the number and levels of connections in nature maintain core processes of primary production and decomposition that, in turn, determine how much total biomass a location can sustain.

And in these ecosystems, all organisms are inherently well. An unhealthy gazelle ends up in the jaws of a cheetah. Predation is an excellent mechanism to weed out the ill and sick. Similarly, healthy ecosystems are free of pollutants and primarily free of novel organisms that humans would term pests and diseases. Nature has no concept of 'meaning' in the human sense of individual purpose or health, but it defies entropy to make more. And it does this all the time irrespective of the disruptions. This meaning of resilience is similar to the one explored on the couch.

Resilience in ecology refers to the ability of a system to continue to deliver the same ecosystem services despite disturbance. It bounces back.

Cheetahs are the fastest land mammals, but they also have an eye for the slow gazelle.

Resistance is about not bending to pressure and standing firm. Interestingly, our shrink might use the term **psychological resistance** and see too much of it as a problem for your well-being. People with resistance show combinations of behaviours that include perfectionism, criticising, disrespectful attitude, self-criticism, preoccupation with appearance, social withdrawal, need to be seen as independent and invulnerable, or an inability to accept compliments or constructive criticism, not exactly what is said about resilience. The reason is that resistance is the ability to ignore or bat away trauma. A suit of armour offers resistance to the lance. A boxer in the ring who keeps getting hit but doesn't fall over shows resistance and an iron jaw.

Ecologists refer to resistance as the ability of an ecosystem to remain unchanged when subjected to disturbances. English ecologist Charles Elton used the word resistance to describe environmental traits that limit the capacity of introduced species to occupy ecosystems successfully. These characteristics include abiotic conditions like temperature and drought and biotic factors like competition, parasitism, predation, and a lack of required mutualists. Resistance can also be influenced by increased species variety and decreased resource availability. In short, disturbances have no detectable effect on the ecosystem because the organisms resist change, as the boxer must to stay in the fight.

Resilience + resistance = stability

The pugilist, who is flawed by a solid right hook, gets knocked down and gets back up again, as the song goes, shows resilience. He takes the hit and returns for the next one. But winning the bout is made more likely by resisting and not getting knocked down in the first place.

Psychologists see resilience as good and resistance as not so good, even though they are both necessary for mental stability. Too much resistance and a person reduces their resilience, especially their capacity for connection; not enough, and we spin out at the smallest slight. In ecology, resistance is the ability of nature to stand up against disturbance and resilience is the ability of organisms and processes to bounce back. Together they confer stability on ecosystems.

Stability (of the ecosystem) refers to the capability of a natural system to apply self—regulating mechanisms to return to a steady state after an outside disturbance.

OECD Glossary of terms

Some ecosystems show high resilience and relatively low resistance. For example, many eucalyptus forests in Australia are impacted periodically and unpredictably by wildfires that are frequent enough to occur within the lifetime of trees. Burnt trees can recover using stored energy and nutrients in the roots to sprout new branches along the trunk, a process called epicormic growth. In a few months, biomass production is up again, and the system returns to pre-fire capability.

Meanwhile, a lake in northern climes can freeze over in winter, the organisms resisting the cold directly or altering the life cycles to miss the impact and emerge again in the spring. We assume that if ecological stability is tested, nature either resists or bends and bounces back. If we think of nature as having resilience at some level, it justifies our resource use. Resilience thinking can replace the old way of describing the landscape as environmental assets for human use. It also helps protect the most valuable assets, including the reserve system and laser-levelled cotton fields.

Resilience and resistance are properties of nature. Together, they can make ecosystems look stable and behave predictably even when buffeted.

Just as the psychologist likes you to build resilience more than resistance because the latter can slip into unwanted responses, we also like ecological resilience. If ecosystems are resilient, we can push them around and not worry—they will bounce back. But here is the thing.

Ecological resilience is a consequence of diversity. It emerges because there are many connections between organisms, and many types of organisms are doing the connecting. Human food production intensifies through simplification—we grow more food in simplified ecological systems. We call them fields, paddocks, meadows, gardens and farmland. So, if we want resilient systems to grow food, they have to become more diverse, not less. Alternatively, we can gamble that they resist everything we throw at them.

Key points

- A key driver of broad-scale land clearing that can lead to environmental degradation is that it is often more profitable to clear new land and mine the natural capital of the soil than to invest in land that has been under production for many years.

- Resilience and resistance are important for stability in food production. Equally important is building mental and emotional resilience in farmers, enabling them to adapt well in the face of adversity.

- The food supply chain is optimized to maximize profit, focusing on efficiency and transfer rather than stability and resilience. One consequence is limited storage capability that forces high turnover in warehouses and retailers, making the system insecure. The COVID-19 pandemic highlighted the fragility of the food supply chain, with panic buying leading to shortages of certain items.

- Integrating food production into nature involves compromise and dealing with native herbivores that can cause damage to crops.
- Super-efficiency and consolidation can lead to a single entity dominating the market, which poses a risk to the resilience of the food supply. The chicken production industry in Australia is a good example.
- Achieving sustainable food production requires structural changes and incentives to balance production and ecology. Persuading people to do things differently and recognizing the importance of both technical and pragmatic solutions is crucial.

In the next chapter

We now know that food production is an energy sink that we assume is resilient, resistant, and stable. History backs us up, but what about the future? What are some of the consequences of intensifying agriculture for growing enough food?

Chapter Six

Intensifying agriculture has consequences

Energy source to energy sink

Whilst we like to quote from scientific sources, they are usually of the pithy variety to illustrate a specific point. This one from a review by Spanish researchers is stodgy, but it is worth reading in full because it is the best summary of what has happened to the land since humans started their population explosion. It is 210 words, which should take you a minute to read. References were removed to shorten the quote, and C refers to carbon.

> *Global cropland area expanded about 4-fold since 1700, up to ca. 20% of the vegetated area. In parallel, agroecosystems intensified to meet increases in population density, culminating in a radical socio-metabolic transformation, with*

the transition from traditional organic, solar-based systems to industrial, fossil fuel-based systems. The former usually operated at the local scale, relying on solar fluxes and internal biomass recycling as sources of energy and fertility. The latter are intensified through imports of fossil fuel-based industrial inputs and international or interregional trade to optimize the conditions for commodity production in a context of a global market economy. These structural characteristics shape the C cycle through effects on the type and quantity of soil C inputs and on the biotic and abiotic factors controlling C losses. Soil C inputs can potentially increase with industrialization due to a higher overall biomass production and a lower use of crop residues for animal feeding. Modern crops, however, usually have higher harvest indices, which reduces the production of residue relative to the main product. In addition, weed biomass is more effectively suppressed in modern cropping systems, and root growth in relation to aerial biomass is usually reduced. SOC mineralization is also affected by the changes in management practices such as tillage, irrigation and fertilisation.

Aguilera, E., Guzmán, G. I., Álvaro-Fuentes, J., Infante-Amate, J., García-Ruiz, R., Carranza-Gallego, G., ... & de Molina, M. G. (2018). A historical perspective on soil organic carbon in Mediterranean cropland (Spain, 1900–2008). Science of the Total Environment, 621, 634-648.

Let's try to summarise this explanation into a concise version. Agriculture went from an energy source to an energy sink. Organic, solar-based food systems changed to industrial, high-yield, fossil fuel-based systems with an international supply chain driven by a need for efficiency motivated by profit. And with this change, the soil gets fewer organic inputs and loses SOC over time. I cannot stress enough the importance of that last sentence. The loss of soil organic carbon and the potential to recover soil organic carbon is the topic of the *Mindful Sceptic Guide to The Soil Carbon Debt*, so the point here is the magnitude and consequence of the expansion and the change from organic to industrial food production. It is as significant a change as a '*Don't look up*' asteroid hitting the earth. Not as immediately catastrophic but just as destructive. However, there is a crucial difference.

Like in the movie, humanity can do nothing about a giant asteroid on a collision course. Deflecting the path is beyond our current technology and skill—no matter what NASA claims about their recent asteroid nudge. However, we can do something about our over-reliance on fossil fuel-based agriculture and the decline of soil organic carbon. We can return to a more organic-based system powered only by the sun's energy and adopt a wide range of production practices that add more organic matter to soil than is lost from land management and mineralisation.

Read that 210-word quote again. Slowly.

It is a wonderfully concise summary of what has happened to agricultural production over the last 100 years and the critical consequences of those changes. We know what happened to agricultural production and can determine how to reverse some consequences without risking food security. And that is well worth your attention. Then, latch onto the hope that such knowledge imparts. But before that option, let's review just how far down the intensification path the global food supply chain has gone.

Energy-blind food supply and the agriculture energy sink

What do you make of this graph from the fascinating book '*The Future is Rural*' by biologist and farmer Jason Bradford, which tells the story of food production in the US in the currency of energy? Start on the right side of the graph with the food energy output per year as food consumed by the human population of the world, 1.1 Quads, where a Quad is a measure of energy. Now compare that to the energy used to generate that food energy eaten—it takes 14.2 Quads to grow, store, deliver, sell and cook the food eaten.

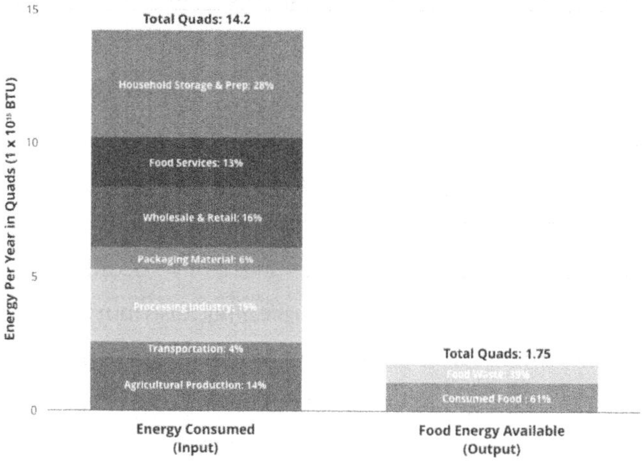

A Quad is 1,000 trillion BTUs and one BTU is the heat required to raise the temperature of one pound of water by one degree Fahrenheit.

This ratio is not sustainable. It's bonkers. Recall that the logic behind the invention of agriculture 12,000 years ago was that it generated an energy surplus, more energy out than labour energy in, and enough food energy to get people through lean times and seasonal shortages of game and forage. The energy surplus allowed people to stay put with a division of labour and create complex societies.

Modern food supply systems are the opposite. They create a colossal agriculture energy sink. Here is how Jason Bradford summarised the numbers in the graph.

> *The modern food system runs a large energy deficit and so do most farms. Our energy-sink food system is an aberration that will eventually appear as a blip in human history, and*

we will require our farms to run an energy surplus to avoid starvation.

Jason Bradford

Agriculture used to produce excess energy. Modern food supply chains run as an unsustainable agriculture energy sink. And it is a huge sink. But why is the agricultural energy sink a blip? Because the energy sunk into the production, processing, packaging, distribution, and waste of food comes from fossil fuels—an army of energy slaves that we dig up and pump out of the ground for the tiny cost of extraction.

And that energy source is about to run out in the case of oil, whilst coal and gas we should not burn or risk massive climate shifts. We are lounging in a gluttonous age of cheap energy and cheap food unprecedented in human history and converting the gift into people.

And before we get all fired up and look for someone to blame, relax. This is neither good nor bad; it is just ecology. Give any organism excess resources without controls, and the population will expand at the maximum reproductive rate. Humans use fossil fuel energy to create an excess of food that we convert into more humans, some six billion more, since oil became the primary power source. Growth is what nature does with an energy source. It uses the power to defy entropy and make more.

Humans are no different to all other organisms in this basic biology—we make more. Our exception is the ability to capture and use exogenous energy (energy from outside ourselves) and apply that energy through technology to change the environment to our benefit.

And we do something else that is different. We are energy blind.

The market delivers the agricultural energy sink

Nobody can do business at a loss, so the push for profit encourages efficiency. Such is the claim of industrial agriculture and the supply chain that maintains it—big fields sown with monoculture crops. If this claim is valid, how can the food supply system use 13x more energy than gets eaten?

We are told that the market is efficient, but the food supply systems only function because of the cheap energy from fossil fuels. The food supply chain absorbs the energy cost needed to function. In other words, the profits that accrue to the businesses in the food supply chain and, by extension, the businesses that rely on people who exist because of the food, are subsidised by ancient energy—the hard work of plants and animals millions of years ago. And this is how we determine efficiency for all businesses. We measure their efficiency in profit to shareholders and are blind to their energy efficiency because energy is ubiquitous and mostly cheap. The only time energy is considered is when a company has a high operation cost, so a few lights are converted to LEDs. Most businesses are assessed as efficient from a profit perspective rather than from collective energy use. Capitalism chooses to be blind to its energy use. What humans do differently from nature is that nature is energy efficient—modern humans are not.

Don't blame the farmers

Food production, the base of the supply chain that comes from the efforts of farmers, uses shy of 2 Quads for the 1.1 Quads eaten. It is not an ideal net efficiency, but it is out of the park compared to the 12.2 Quads needed to make the food energy available to the household or the 0.7 Quads of food waste at the table.

Tempting as it is to blame the farmers, they are not the problem. The real problem and the opportunity for substantial energy savings is consumer behaviour—what we choose to eat, how we prepare our food and what we manage to waste. But even the people are not to blame. Most of us consume what we can readily access. We eat the food we can afford, the food in the supermarket, and the food we can prepare easily after a long day at the office.

So, if you want to blame someone, try the food supply chain operators or, more strictly, the people who let market mechanisms loose on our food systems. And blame them for their energy blindness. But even here, blame is not warranted. How many of us rejected the food, health, education and comfort generated by capitalism? I didn't.

Energy-blind food supply chains

Food systems use energy in the production, storage, processing, packaging, distribution, consumption and disposal of food. Jason Bradford's graphic tells us that these energy inputs are 8x greater than the food energy consumed by people. And this has not mattered. Businesses in the supply chain accept energy bills as a cost of doing business, recoverable in the price of food at the checkout.

Smart companies look to energy efficiency to increase profit margins, but cheap energy availability means they fail to see the more significant energy consequences. The food supply chain works even with an 8:1 energy sink ratio. And they will stay energy-blind.

If they opened their eyes, they would realise that the energy-intensive production and distribution systems are not sustainable and will collapse when oil becomes scarce. Many will go out of business through whatever energy transition humans invent. No CEO would admit this existential risk to the company shareholders. Better to keep all eyes shut.

Fortunately, farmers are better placed to cope. They can revert to less energy-intensive forms of production as the costs of inputs rise. Recall that the pre-industrial farming systems generated an energy surplus because nutrients and carbon were recycled, and energy inputs came from human and animal labour. But can retro-agriculture grow enough food?

Clean energy solutions

Mechanisation and globally traded commodities facilitated by fossil energy have steadily replaced human and animal labour and local markets in the industrial food system. Every step in the supply chain requires energy and lots of it—a person eating a meal in the United States consumes only one kilocalorie of food for every 10 kilocalories spent getting that food to their plate.

It may be that a new energy source emerges that doesn't pollute the planet. There is already a rush on renewables that need more resources than the planet can cough up just to build and maintain, and the technofix may lull some people to sleep. Clean energy would be tremendous.

Sustainable clean energy would be magnificent. We don't have either, not at scale.

And even if the miracle of fusion, hydrogen or some other clean source of cheap energy emerged at scale, the food supply chain faces a complex structural transition to electricity to replace diesel-powered machines and gas-fired fertilizer factories. Assuming this challenge is met with EVs and factories run on fusion power transition, still growing enough food will demand our attention. Then there are the other structural risks, not least the soil organic carbon debt and ultra-processed foods delivering an epidemic of metabolic illness. Growing enough food is just the start. But what about alternative production methods? Surely, they can help.

Feeding people by farming in air

Soil is the primary medium that gave humanity its leg up into the modern world of vast food production and billions of people. Soil took the windfall of cheap fossil fuel energy to support biomass as crops and livestock often much more than would occur naturally. Secure food production is why land is so valuable, and the control of land is central to the integrity of society and those who rule. Soil supports food production that delivers our calories and nutrients. Water is important, too, for irrigation and for fish that provide about 3.3 billion people with almost 20% of their average per capita intake of animal protein and 7% of global protein consumption. But it's the land where we grow most of our food.

All around the world, agricultural soils are under strain. Here are some estimates of the extent and impact of global land degradation from a report to the Global Environment Facility on sustainable land management.

- *About 24% of total global land area has been affected by land degradation, including over half the total farmed area*

- *Globally 1.5 billion people and 42% of the very poor are directly affected by land degradation*

- *Each year an estimated 25 to 40 billion tons of fertile soil are lost globally*

- *In the drylands, due to drought and desertification, 12 million hectares of land are transformed into new man-made deserts each year. That is an area with potential to produce 20 million tons of grain every year.*

- *In the last two decades, significantly land recovery and Improvement have occurred in the drylands. For instance, many regions have already adopted farmer-managed natural regeneration and agroforestry techniques, such as planting of 'fertilizer trees' on farmlands and grazing lands. Such techniques have contributed to improving over 6 million hectares across Africa but more than 2 billion hectares of land worldwide is suitable for rehabilitation through forest and landscape restoration.*

- *In all 67% of Africa's land is affected by land degradation, 4 to 7% of land in sub-saharan Africa is severely degraded.*

- *The cumulative loss of productivity is 25% of cropland and 6. of pasture land.*

- *Half of soil degradation is attributed to overgrazing.*

Source: Henry, B., Murphy, B., & Cowie, A. (2018). Sustainable land management for environmental benefits and food security. A Synthesis Report for the GEF (Global Environmental Facility). GEF, Washington, DC, USA. Acronyms: FAO, UN Food and Agriculture Organisation; ITPS, Intergovernmental Technical Panel on Soils; UNCCD United Nations Convention to Combat Desertification.

These numbers refer to land degradation, but they mean that 40% of global agricultural soil is losing the capability to grow food. The UNCCD refers to land degradation as the reduction or loss of biological or economic productivity and complexity of the land resulting from natural processes, land uses or other human activities. More formally,

> *The reduction or loss, in arid, semi-arid and dry sub-humid areas, of the biological or economic productivity and complexity of rainfed cropland, irrigated cropland, or range-land, pasture, forest and woodlands resulting from land uses or from a process or combination of processes, including processes arising from human activities and habitation patterns, such as: (1) soil erosion caused by wind and/or water; (2) deterioration of the physical, chemical and biological or economic properties of soil; and (3) long-term loss of natural vegetation*
>
> United Nations Convention to Combat Desertification (UNCCD)

Land is a more accessible term than soil. The land is mapped and legally owned and traded. It can be fenced and managed. A person can go and stand on a plot of land and look at it. But the reality is that degradation refers to the properties of soil and the consequences of changes to these properties for vegetation. These land degradation numbers mean that a significant and growing proportion of global soil used to grow food produces less biomass than it could under given conditions. An anthropomorphism would be to think of land degradation as unhealthy soil, not as fit or as strong as it could be.

Loss of soil health is most serious for the poor. It is acute in places where the pressure on land is increasing rapidly, particularly in Africa and South Asia.

Growing pressure on unhealthy soil

At the current population growth rates, a country like Niger in West Africa, with 26 million people in 2022, is expected to have 65 million by 2050, more than double in a generation, and, all things being equal, reach 100 million by 2068.

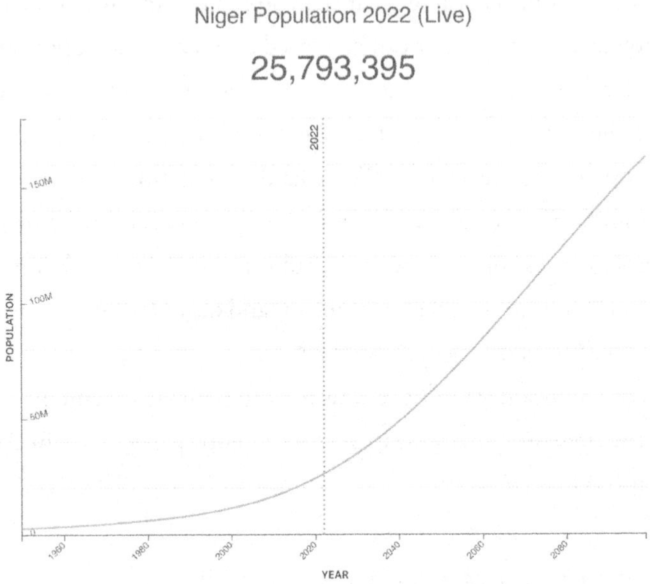

Population growth in Niger as an example of what is expected to happen in Africa over the next 50 years; lots more people. Source: World Population Review

Soil cannot grow food at the rates such increase demands. Resource limits will likely impact the shape of these population curves, and all things will be far from equal. But until that restriction happens, alternatives to soil are critical to maintaining food supply and helping achieve food security, especially for the poor.

Arable land in Niger is around 17 million ha, and there is only a small area left with soil suitable for arable expansion. Without any food imports or crop production, the current cropping area would need to support 1.5 persons per hectare in 2022, 3.8 persons per hectare by 2050, and 5.8 by 2068. This is still below the global average of 6 people fed ha^{-1}, but Niger is semi-arid and, even without soil degradation, cannot match the production average. For example, maize yield in Niger is around 1 t ha^{-1}, close to the global average but half the yield of US farms.

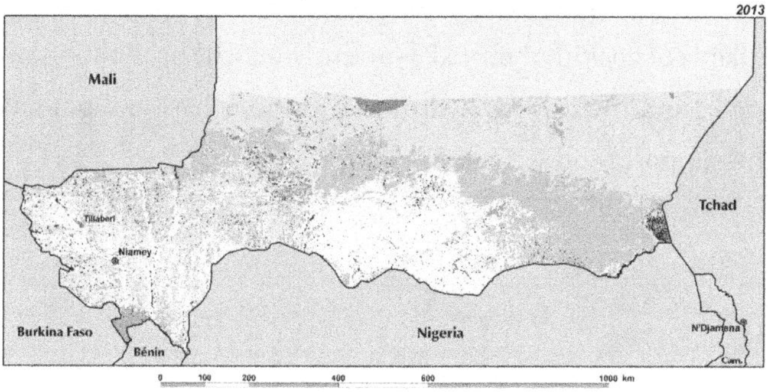

Pattern of land use in Niger. Source: USGA Land use and land cover trends in Niger. Yellow represents arable land, green is Sahelian short-grass savanna and orange is steppe

At 4% population growth, dry climate, and limited scope for expansion of arable land with traditional agricultural methods, a country like Niger is unlikely to be self-sufficient in food from soil alone. Even with careful land-use decisions, best practice land management, and readily accessible fossil fuel inputs, the production systems in a dry country on the edge of the Sahara are too fragile—yield gaps are already significant, water is limiting, and soils are degraded. Meeting the demand for calories would require imports unless alternative production methods can be found. So what could be done?

Farming in water and air

Technology is helping grow food without soil. Water-based systems to grow plants (hydroponics) that deliver nutrients to plant roots via solutions have been around for some time. Plants grown in water are great for leafy greens, tomatoes, some vegetables and strawberries.

Combined with fish (aquaponics), neat, partially closed-loop systems are possible where the fish provide nutrients to the plants and a tasty grilled tilapia to go with the salad. Currently, a niche production system, aquaponics and the various hydroponic options have the potential to scale and support food security.

Aquaponic systems for strawberry production decrease the need for external nutrient supply but also reduce the overall water discharge and increase water use efficiency.

Other innovations have dispensed with solutions altogether to grow plants without soil or water. Plant roots dangle in the air and are moistened periodically by specially designed misting devices (aeroponics). This allows growers to raise crops in open-air, misty environments or mineral nutrient solutions under indoor controlled conditions instead of soil.

I know it's a bit out of the box, but farming in the water or the sky can augment food production from the soil. There are huge implementation challenges, especially scaling aquaponics or aeroponics to feed whole communities of people. It's a hard problem but not impossible.

A key advantage is that the growing population is looking to find reliable food sources and ways to engage in food production. Hungry people will try anything and, given a chance, will apply their intellect and enthusiasm to new opportunities. The question to ask about these technology fixes and novel solutions is this.

Where are the nutrients, energy and materials to build and run novel infrastructure?

Even the smartest innovations are a net transfer of nutrients and energy from nature into food for people. The energy and nutrients must come from somewhere, and there must be enough water to lubricate the process. Innovation is possible and will happen. Humans have always done it, but there is no silver bullet in any of the 'ponics' because they still need inputs. The truth is that soil remains the substrate on which future food will be grown and raised. And despite the degradation, there is hope because we know so much already.

We already know what must be done for food security

We know that food production has to happen to feed everyone and that farming practices create their own landscapes, soils and rural culture. Most food is grown in fields away from urban areas, so people in town (over half the global population) rely on a food supply chain for their basic needs.

We know that around a third of current and historical greenhouse gas emissions come from agricultural practices and clearing land for agriculture. There are historical emissions and pressure for more, given that food production has to keep going to meet demand.

We know that the cleared land and the farming landscapes are not where biodiversity wants to hang out. As habitat is restricted and fragmented, biodiversity loss is inevitable. At the same time, rural landscapes are increasingly hostile to biodiversity, including critical groups like insects and microbes.

We know that current food production systems and supply chains are fragile and susceptible to predictable and unpredictable risks. Industrial-scale intensification meets the demand for food production through high input of energy and chemicals that easily collapse if the inputs stop.

But what we know is not all bad news. Thanks to proven, science-based ways to create more efficient, less polluting and more resilient agricultural landscapes with less mining of soil, our prescience tells us there are many food ecology actions we should take, for example:

- keep vegetation cover on the soil at all times and reduce tillage to a minimum

- protect patches of remnant native vegetation — habitat for animals and helps draw down and store carbon from the atmosphere

- retain water in the landscape by slowing the overland flow and creating healthy farm dams which can provide higher-quality drinking water for livestock, improve farm productivity and create wildlife habitat

- plant "shelterbelts" – strips of woody vegetation that shelter livestock from wind and sun, provide wildlife habitat (when well designed and managed), and prevent moisture loss from soil

- control weeds and pests
- shorten supply chains wherever possible
- reduce waste so that the majority of food produced gets eaten

The most important is maintaining as many of these actions as possible and simultaneously because no single solution works. It must be integrated.

We know the problem—fragility, risk, and unprecedented demand. We know the solutions—integrate multiple actions that maintain healthy soil, reduce waste, and use more ecology to reduce inputs. There is nothing in future food security that we cannot see coming, and we already have most of the evidence to deliver and justify actions to mitigate risk.

Fixing the broken food system

Regarding food system priorities, we usually default to the human dimension, perhaps through the four elements of food security—availability, access, utilisation and stability. Food is for people, after all. But check out this opening stanza from a science article in the learned journal Nature, perhaps the most prestigious for serious science about the natural world.

Land use and food production are not meeting people's needs. Agriculture destroys forests and biodiversity, squanders water and releases one-quarter of global greenhouse-gas emissions. Yet one-third of food is wasted, 800 million people remain undernourished, 2 billion are deficient in micronutrients, and obesity is on the rise.

Guido Schmidt-Traub, Michael Obersteiner and Aline Mosnier (2019) Fix the broken food system in three steps Nature 569, 181-183

We'll overlook the emotional, normative verbs in the passage (destroys, squanders, wasted) but take the overall message—**land use and loss of biodiversity present a major risk to food production, as does the human tendency for profligacy in anything we consume**.

And there is the inequality statistic. If over 10% of the human population lacks food, how can anyone with any humanity condone so much waste? Remember that 10% of the global population is over 800 million people.

Most of us in the West are aware of the obesity issue, which is as much about a change in diet to sugar, grains and seed oils as it is about sedentary lifestyles. Our bodies did not evolve to handle either. The micronutrient one is the sleeper. Nutrient deficiency could lead to diseases and vulnerabilities for want of the elements that we need only a small amount of but are essential to so much of our bodily functions. Check out Chris Scott's *Mindful Sceptic Guide to Sustainable Diet* for a deeper dive into diet.

As we have seen, the rhetoric is that the six-continent supply chain is an economic powerhouse. It soaks up capital and adds labour from people and fossils to generate returns in dollars and more people. Only the profit focus is not sustainable. Through its lauded efficiency, it mines natural resources and exports waste products to the environment. It has worked to get us here, but it has failed to meet everyone's needs.

The perspective from the scientists is that the environment and especially the soil and agricultural landscapes that produce our food, are what we have to meet human needs. Shifting from person to place for food system priorities can help to understand the challenge.

European researchers Guido Schmidt-Traub, Michael Obersteiner and Aline Mosnier propose three food system pillars as a solution: 1) efficient and resilient agriculture systems, 2) conservation and restoration of biodiversity, and 3) food security and healthy diets. Loosely, ecosystem services, conservation, and food sense are linked because they rely on nature. Schmidt-Traub *et al.* also propose that these pillars are consistent with sustainable development—a bit of an oxymoron, but we will discuss that another time. The solutions are also about how we use nature, what parts we want to keep, and our sense of well-being. This is true even though the issues point to the very base of our hierarchy of needs.

Let's take a closer look at these pillars that help us shift from person to place.

Pillar 1—Efficient and resilient agriculture systems

The market has made food production efficient. Priorities for efficiency and greater production are good, especially for the businessman or investor—nothing like a profit imperative to minimise costs and maximise returns. The real challenge is resilience. And not just because profit imperatives tend to externalise costs onto the environment as pollution and can tolerate food waste through economies of scale and the consequences of competition. The real threat to resilience is intensification, the shift in agriculture from an energy source to an energy sink.

Pillar 2—Conservation and restoration of biodiversity

Almost all food is grown and reared. In other words, applied ecology. Agricultural production systems have simplified the ecology to focus on a handful of plant and animal species that comprise the bulk of human food. But what food ecology tells us is that these simpler systems are not only less diverse biologically, but they are also more fragile. There are fewer species to pick up the slack if one or more are lost.

Conservation also makes sense for the collective ecosystem services that biodiverse systems deliver. Holding water in a landscape with plants to cover the soil is easier. And where the biodiversity matters is in the soil. Pillar 2 solutions for protecting and preserving nature make sense but are easily at odds with Pillar 1 objectives.

Pillar 3—Food security and healthy diets

Zero hunger, low dietary disease, and reduced food waste are goals borrowed somewhat from the UN Sustainable Development Goals. How hunger, dietary disease, and food waste are to be reduced is not suggested. The *Mindful Sceptic Guide to Sustainable Diet* combines the need for food with the need for metabolic health through nutrition. It links it to the food system through what a person eats, what she feeds her family and the collective diet of nations and the human species. Again, that is understandable, for not only are these technically challenging, but they have to deal with human decision-making. Bring psychology into anything, and not only is it closer to the guts of the issue, but it also makes it wicked.

Focusing on food system priorities?

Do you remember the Conference of the Parties to the Convention on Biological Diversity and the United Nations Framework Convention on Climate Change? Probably not. If you are under 25 years old, it was before your time. The UNFCCC might be familiar from all the press on climate change, but don't confuse it with the IPCC, which predates UNFCCC by four years. The IPCC is the panel of scientists who tell us what's happening with the climate, and the UNFCCC is the international agreement they helped create. The COPs, a conference of the parties, including the famous COP 21 in Paris, come around towards the end of each year and sit under the convention. But I digress.

The UN conference that delivered seminal global environmental statements went down at the infamous Rio Convention in 1992. For the first time in history, the environment attracted world leaders, prime ministers, and presidents instead of just environment ministers who didn't even have a corner office. These UN proceedings and declarations were supposed to be both a statement of the problem and a call to action for biodiversity and climate. We will outline what happened elsewhere, but one word is enough for this discussion—failure.

A big, fat, undeniable failure. Even after 27 COPs, the climate continues to change with increasingly dire consequences for ecosystem goods and services. Biodiversity loss is still on a downward trajectory. And yes, the UN Convention on Biodiversity has COPs, too; COP 15 occurred in Montreal, Canada, in December 2022.

Meanwhile, the drivers of degradation are rising: human land use, energy use, and economic activity. In other words, the solutions to building and maintaining food system priorities are difficult to find and deliver. Internationally, we can't agree, and we are beset with detail locally. Still, the global call to action that happened in 1992 at the Rio Convention is just as important today, more so if anything. There are more people and fewer resources than there were 25 years ago. Over 2 billion more people. Or, if you prefer, a 42% increase in the global population of 5.5 billion people alive in 1992, the year Madonna released her *Erotica* album, and Whitney Huston was number one in the singles charts with *I Will Always Love You*. There were 2.3 billion people alive at the time of World War 2.

Solutions to the increasingly acute problems humanity faces are not of the silver bullet kind. They cannot be fired at one target because there is no one target—one challenge affects another that links to another and then a fourth.

But here is the hope.

Nature has made it work for billions of years, so we know it can be done. Before modern humans proliferated, global biodiversity was as diverse as it had been in evolutionary history despite the past mass extinction events. There were more than enough resources to allow a single species to grow in numbers and colonise every continent. In a crude sense, we must figure out how nature did it and do it again.

Thank you to Guido Schmidt-Traub, Michael Obersteiner and Aline Mosnier for their framework of food system priorities—efficient and resilient agriculture systems, conservation and restoration of biodiversity, food security and healthy diets. I also thank them for their premise that the food system is broken and needs to be fixed. Profit-based production feeds people but at the cost of disconnecting from nature. The food system's priorities are to return to how food is grown, processed and eaten... and still feed everyone well.

It is tempting to make this a 'back to nature' option that has everyone on a paleo diet, and half the planet rewilded. Only there is a snag—8 billion people—7% of all humans that have ever lived are alive today and need food. And so we come to the critical caveats.

Understand that resources are finite and trade-offs in resource use are inevitable.

If we want more food, there will be a cost to other environmental values. The challenge is understanding the specifics of the trade-off and deciding which resource to use, when and by how much so that different values can be maintained.

It is not possible to retain all values everywhere. I am sorry, but this is the truth.

It is possible to be smart about the choices, though. Indeed, being smarter starts with recognising there is a trade going on and that there are linkages between the priorities; they are hard-wired together. This linking of the pillars is the key—they must be connected through their foundations.

Key Points

- Agriculture has transitioned from being an energy source, relying on organic, solar-based systems, to an energy sink, driven by industrial, fossil fuel-based methods with international supply chains. This shift has had significant consequences for soil health and carbon content and is unsustainable.

- Around 40% of agricultural soil is losing its capability to support food production. This soil degradation poses a significant challenge to food security, especially in regions with rapidly increasing populations.

- Alternative methods of food production, including hydroponics and aquaponics, which use water-based systems to grow crops, and aeroponics, where plants are grown without soil or water, relying on specially designed misting devices are options for growing more food. However, even innovative approaches are a net transfer of energy and nutrients from nature into food, and securing these resources is crucial.

- Three food system priorities: 1) efficient and resilient agriculture systems, 2) conservation and restoration of biodiversity, and 3) food security with healthy diets could help address the inefficiencies and environmental impacts of the current food system.

- The complexity of environmental challenges and their interconnections mean there are trade-offs and choices in resource allocation to balance competing values. Solutions to these challenges require understanding these interconnections and making choices that balance competing values.

In the next chapter

Human institutions drive food, ecology and diet from supply chains to governance. Society is also about people, their thinking and behaviour, and their emotional responses. Feeding everyone needs society.

Chapter Seven

Growing enough food

Humans are unique in the animal kingdom. We have invented and gained much of our success through cooperation beyond our kin that is sustained through what we call society, *the aggregate of people living together in a more or less ordered community.*

Society gives us trade, culture, commerce, collaboration, and mutually assured destruction. It is held together with rules and institutions invented by people, and for the most part, we agree to and allow governance over our lives. We do this because it is much better than the free-for-all that would happen if we scrambled for resources. Other species communicate beyond family groups and for close kinship bonds, but none have overcome genetic competition for mutual benefit. This should give you hope.

Society is both a constraint and an opportunity for growing enough food for feeding everyone. How society, with all its rules, norms, inertia and diversity, chooses to use natural capital will determine how much food can be grown, the quality of that food, equity in food distribution, and, ultimately, the likelihood of human survival into the next century. Growing enough food is a people problem solved by people.

What is a good society?

A good society is an opinion, meaning there are many good ones without an objective way to decide the best. Good would usually mean that most people were satisfied with their lives, believed that they had an opportunity to better themselves, that their children would do better than they did, and that people could walk down the street without getting mugged. Implicit is that there will be more.

In practical terms, this comes down to food, water, shelter, safety, health care, education, and the ability to earn a living. Currently, trading time for a realistic wage for the majority.

Good would also include a feeling of self-determination. Society was in the background, supporting and nurturing but without overt control. Most of the time politics would be about debate and not interference. Society can deliver these things in a multitude of ways, from the draconian to the progressive and in recent decades, we have come to believe the democratic model is the one that achieves these outcomes most consistently.

When it comes to feeding everyone, a 'good society' for growing enough food is a **culture and governance that sees the food security challenge, believes in equity, recognises multiple values, and understands that we rely on nature for wellbeing.** In short, a good society is about sustainability.

Growing enough food

The invention of agriculture just 12,000 years ago, transformed human societies, enabling them to thrive in large, settled communities. The Industrial Revolution, driven by the exogenous energy from fossils and technologies, transformed agriculture into an energy-intensive industry and the growth of food production to feed billions of people worldwide. Whilst the intensive harvest feeds most urban dwellers, 500 million farmers still grow food for subsistence with minimal inputs.

Modern food production and supply chains challenge ecological limits, depend on nutrient and energy inputs, and cannot function without fossil fuels. This is not sustainable and will fail unless there is a transition toward more ecological and sustainable farming practices, such as circular agriculture, to ensure long-term food security and environmental preservation.

Humans grow enough food, but food security means adequate availability, access, utilization, and stability of nutritious foods. Close to a billion people face food insecurity despite enough food production to meet demand. Ensuring everyone has access to nutritious, affordable, and consistent food supplies is a complex global challenge.

The reality is that modern food production is an energy sink despite being designed to generate energy that humans respire—food. Such a system is inherently fragile and unstable. Not what you want when there are 8 billion mouths to feed, increasing at 8,000 an hour.

Society is complex, dynamic and often poorly understood. People bring their prejudices, fears and opinions before they find compromise and solutions. This produces inertia, suggesting that society is slow to change but can be disrupted instantly when enough people decide that change is necessary.

It is impossible to feed everyone without understanding how society works. Human institutions, from money to government to the family unit, impact individuals' decisions as they live. Remember, though, that society and its institutions are a human construct. As such, they can be taken down and replaced. And they might need to be.

As I write, there are roughly 550 million farmers worldwide who collectively grow enough food to feed everyone. We know this because a third of production is wasted along the supply chain, grains are fed to livestock, and crops are grown for biofuel. There are inequality and access issues that mean over a billion people are always hungry, and yet we grow enough food for now. Tomorrow is another day. Unfortunately, a mindful sceptic may lose sleep getting there.

Scence sources

All the *Mindful Sceptic Guides* are compiled from decades of experience with the science literature, but I also gathered specific *Evidence Reviews* that you can access by clicking on or scanning the image below.

Mindful Sceptics

evidence reviews

scan or click the QR code for a deeper dive into the science sources for the Mindful Sceptic Guides

SCAN ME

About the Authors

Mark is an ecologist, not a green one. He has built, run and sold companies, held academic posts at four universities, advised governments on environmental matters, has a PhD and still plays too much golf. His ecological research ended up in more than 70 peer-reviewed research papers and over 80 technical reports, and he supplied peer-review for more than a dozen international science journals. He lived in Africa for a decade, where he researched termites and millipedes, and once walked into a lion who snarled at him. Much of his career has been in Australia, where he won an Outstanding Teacher award from Macquarie University.

His consultancy practice advises commercial and public sector clients on assessing, monitoring, and evaluating environmental impacts, especially in carbon asset and liability management. His non-fiction books are about awareness of humans' role in nature and, as Paul Sorol, he chronicles and writes fiction for a change.

Link in bio

Chris is an ecologist, a restoration one. A lifelong mindful sceptic, Chris has happily landed as an ecological practitioner in his bush regeneration business. When not passionately managing land, being awarded Landcarer of the Year, trawling the evidence on nutrition, diet and health or carefully advising NGOs and government, he grows plants in his commercial nursery.

MINDFUL SCEPTIC GUIDES

Abatement

Biodiversity loss

Escaping the Malthusian trap

Farmers

Food security

Food, people and soil

Growing enough food

Healthy and sustainable diet

Sustainability

The soil carbon debt

mindfulsceptics.info

Made in the USA
Monee, IL
04 November 2026

67379686R00085